T0137037

Emergence, Complexity and Computation

Volume 31

The Emergence, Complexity and Computation (ECC) series publishes new developments, advancements and selected topics in the fields of complexity, computation and emergence. The series focuses on all aspects of reality-based computation approaches from an interdisciplinary point of view especially from applied sciences, biology, physics, or chemistry. It presents new ideas and interdisciplinary insight on the mutual intersection of subareas of computation, complexity and emergence and its impact and limits to any computing based on physical limits (thermodynamic and quantum limits, Bremermann's limit, Seth Lloyd limits…) as well as algorithmic limits (Gödel's proof and its impact on calculation, algorithmic complexity, the Chaitin's Omega number and Kolmogorov complexity, non-traditional calculations like Turing machine process and its consequences,…) and limitations arising in artificial intelligence field. The topics are (but not limited to) membrane computing, DNA computing, immune computing, quantum computing, swarm computing, analogic computing, chaos computing and computing on the edge of chaos, computational aspects of dynamics of complex systems (systems with self-organization, multiagent systems, cellular automata, artificial life,…), emergence of complex systems and its computational aspects, and agent based computation. The main aim of this series it to discuss the above mentioned topics from an interdisciplinary point of view and present new ideas coming from mutual intersection of classical as well as modern methods of computation. Within the scope of the series are monographs, lecture notes, selected contributions from specialized conferences and workshops, special contribution from international experts.

More information about this series at http://www.springer.com/series/10624

Nanda Dulal Jana · Swagatam Das
Jaya Sil

A Metaheuristic Approach to Protein Structure Prediction

Algorithms and Insights from Fitness Landscape Analysis

 Springer

Nanda Dulal Jana
Department of Computer Science
  and Engineering
National Institute of Technology
  Durgapur
Durgapur, West Bengal
India

Jaya Sil
Department of Computer Science
  and Technology
Indian Institute of Engineering Science
  and Technology
Howrah, West Bengal
India

Swagatam Das
Indian Statistical Institute
Electronics and Communication
  Sciences Unit
Kolkata, West Bengal
India

ISSN 2194-7287   ISSN 2194-7295   (electronic)
Emergence, Complexity and Computation
ISBN 978-3-030-09075-3   ISBN 978-3-319-74775-0   (eBook)
https://doi.org/10.1007/978-3-319-74775-0

Printed on acid-free paper

This Springer imprint is published by Springer Nature
The registered company is Springer International Publishing AG
The registered company address is: Gewerbestrasse 11, 6330 Cham, Switzerland

Foreword

Slightly over 50 years ago, pioneering bioinformaticist Margaret Dayhoff published her first Atlas of Protein Sequences and Structure. At the time, the sequences for only 65 proteins were available in the literature, and just a handful of x-ray structures like those of myoglobin or hemoglobin were available. Dayhoffs database led to the development of the Protein Information Resource, which also was the first database that could be accessed by remote computers. Her landmark effort also led to an improved understanding of protein structure and evolution. However, a central problem remained. While it was understood that proteins fold from primary to secondary to tertiary and even quaternary levels, the rules associated with that process remained unknown. Over time and through considerable worldwide experimental effort, the number of protein sequences grew exponentially as did the number of known protein structures. Current databases such as UniProtKB/TrEMBL contain 93,000,000 protein sequences while the Protein Data Bank holds roughly 135,000 known structures. In light of this impressive rapid increase in information, we have learned about structural domains, supersecondary structure, and superdomains. We understand considerably more about the process of folding than ever before. Despite that knowledge, and despite similar exponential gains in computing power over roughly the same time frame, we remain unable to consistently and accurately fold long protein sequences into their native structures via *ab-initio* simulation. We can choose to use supercomputers to calculate all-atom models of the folding process, but these are useful only to a certain time horizon or protein length. We can thread protein sequences onto known structure information through a process of homology modeling to see how well a new sequence maps to a known structure yet this is not *ab-initio* folding. Another promising approach is the use of metaheuristics such as evolutionary computation to iteratively search for improved folds in light of an objective function. But this requires knowledge of the right objective function for the protein and its environment, which may in some cases be difficult to define. This book offered by Nanda Dulal Jana, Swagatam Das, and Jaya Sil focuses on metaheuristic approaches and improved understanding of the fitness landscape that is used as the objective function for search. As our knowledge about sequences, structures, and their mapping improves, we are faced with the realization that proteins in living systems are likely to be in constant motion. Our inability to faithfully recapitulate the correct structure may itself be an artifact of the fuzziness of the actual process, or our inability to capture the fitness function in a manner that makes it possible for either physics-based or metaheuristic approaches to succeed. While we await the exciting decades of experimental protein structural biology ahead of us, we explore and improve our use of computers to help solve these problems. This book by Jana, Das, and Sil continues us on that journey that Dayhoff and others started over 50 years ago.

San Diego, CA, USA   Gary B. Fogel, Ph.D., FIEEE
November 2017   Chief Executive Officer, Natural Selection, Inc.

Preface

Proteins are the main building blocks of the dry mass of cells in living organisms. Proteins perform a huge variety of tasks such as catalyzing certain processes and chemical reactions, transporting molecules to and from the cell, delivering messages, sensing signals, and countless other useful functions which are essential for the preservation of life. Protein is a macromolecule combining with amino acids that are connected by peptide bonds. During synthesis, protein is folded into a three-dimensional structure. However, sometimes protein may fold into an incorrect structure (known as misfolding) leading to a protein substance with the different properties which can be harmful to organisms. Misfolding of a protein causes many critical human diseases such as Alzheimer's disease, cystic fibrosis, amyotrophic lateral sclerosis, cancer, and neurodegenerative disorders. Therefore, the knowledge of protein structure and its prediction by analyzing the amino acid sequences is a challenging task in computational biology.

Biologists define a hierarchy of protein structures with four levels in order to more accurately characterize the structure properties. The *primary structure* is simply the linear protein chain, i.e., a sequence of amino acids. The *secondary structure* is the local conformation of protein structures. There are two types of major secondary structures, known as alpha-helix and beta-sheet. The *tertiary structure* is the global three-dimensional structures. Sometimes, multiple protein chains united together and form hydrogen bonds between each other, resulting in the *quaternary structures*. Therefore, the protein structure prediction (PSP) is the process of determining the three-dimensional structure of a protein from its sequence of amino acids.

The experimental techniques such as X-ray crystallography and nuclear magnetic resonance (NMR) have been used to determine a protein structure from its sequence of amino acids. For instance, it may take months of concerted efforts to crystallize a single protein in order to enable X-ray diffraction methods that determine its 3D structure. The major limitations of experimental techniques are extremely time-consuming and labor-intensive, strict laboratory setup, and heavy operational burdens. As a result, the gap between the protein sequences and the known structures is increasing rapidly day by day. For example, there are 78028152 protein sequences but the structure is known only for 117184 of them as of January 2017. Therefore, researchers from multiple disciplines and with diverse domains of knowledge are involved in predicting protein structure from its sequence of amino acids using computational methods.

Computational approaches have the potential to predict the structure of a protein from its primary sequence in order to overcome the difficulties associated with the experimental approaches. Basically, a computational technique is a process of finding a three-dimensional structure by arranging a sequence of basic structural elements. Depending on the database information, computational methods are divided into three category: *homology* or *comparative*, *threading* or *fold recognition*, and *ab-initio or denovo* PSP. Homology modeling is based on the similarity comparison of the sequence while threading approach is a process to thread together likely short sub-conformation of the corresponding subsequence. These two techniques fully depend on the availability of similar sequence information exists in the database. i.e.. these are template-based methods and the results may become unconvincing for dissimilar sequences. On the other hand, *ab-initio* method predicts three-dimensional structure of a protein from its primary sequence directly based on the intrinsic properties (namely, hydrophobic, and hydrophilic) of amino acids, so it is a template-free approach. The concept of *ab-initio* method is based on the *Anfinsen's thermodynamic hypothesis* where the theory of thermodynamics states that conformation of a protein corresponds to the global minimum free energy surface and the conformation is called a native structure of the protein. A native structure is a stable structure of a protein which always consumes minimum energy among all the conformations of the protein. Therefore, *ab-initio* based PSP approach is transformed into a global optimization problem where energy function would be minimized.

The *ab-initio* based PSP has two important key parts. The first part is to state the simplified mathematical model that corresponds to a protein energy function, i.e., the objective function. The second part is to develop a computationally efficient searching or optimization algorithm which can find the global minimum of the energy function. A number of a computational model have been proposed and developed that derive the protein energy function for the PSP problem. Among these physical models, hydrophobic-polar (HP) lattice model is a simple model widely used for structure prediction. It can be represented in two dimension or three dimension and consumes less computation time, but the level of accuracy of the predicted structure is not enough to use in rational drug designing. The off-lattice model is also a computational model that brings good interactions between amino acid residues and their environments. The *AB* off-lattice is such a type of model where amino acid residues are characterized as hydrophobic (A) and hydrophilic (B) residues. It represents the intramolecular interactions between residues which are connected by rigid unit length bonds, and the angles between the bonds change freely in two or three dimension. So, the *AB* off-lattice model provides a structure of a protein at much higher atomic details than the lattice model. The energy function for the AB off-lattice model is expressed as a weighted sum of bending energy of protein backbone, independent of the sequence. The non-bonded potential energy among nonadjacent residues is known as *Lennard-Jones* 12, 6 potentials. The two components of the energy function are defined by bond and torsional angles between two consecutive amino acids. All angles are limited in an interval $[-\pi, \pi]$. Thus, the objective is to find the optimal angles associated with the energy function that provides the minimal free energy value. In this way, a PSP mission is transformed into a numerical optimization problem. It is widely accepted that *ab-initio* based structure prediction using AB off-lattice model belongs to a class of difficult optimization problem known as *NP*-complete problem. Therefore, in order to solve such problem, search and optimization algorithms are usually developed using certain *heuristic* or *metaheuristic* that, though lacking in strong mathematical foundations, are nevertheless good at reaching an approximate solution in a reasonable amount of time.

Metaheuristics denote a collective name for a range of problem-solving techniques based on principles of biological evolution, swarm intelligence, and physical phenomenon. These techniques are based on iterative progress, such as growth or development in a population of candidate solutions for a given problem. The population is then selected in a guided random search to achieve the desired end. Since the past few decades, algorithms from this domain have attracted the researchers to the field of *ab-initio* based PSP. However, the characteristics or complexity of the optimization problem play an important role in determining the optimization technique to be used for problem-solving and influence the performance of the algorithm. Therefore, a fundamental question arises: how to select and configure a best-suited algorithm for solving the PSP problem? This book attempts to determine the structural features associated with the PSP problem and configured some widely used promising metaheuristic methods based on the structural features to find near-optimal and acceptable solutions. The structural properties are determined using fitness landscape analysis (FLA) which provides a better understanding of the relationship between the problem nature and the algorithm in order to choose the most appropriate algorithm for solving the problem.

Extensive research on metaheuristic algorithms have been proved their potential in solving complex optimization problems like PSP. However, it is not easy to choose the best metaheuristic technique for solving a particular problem. FLA is used for understanding the problem characteristics based on which the best-suited algorithm can be chosen. In the literature, it has been shown that more research on landscape analysis in discrete search space have been considered compared to landscape analysis on continuous search space. Since the PSP problem is defined as a continuous optimization problem, continuous FLA for the PSP problem has remained uncovered till date. In this book, we determine the structural features of the PSP problem and configured some well-known metaheuristic techniques based on the structural properties. Toward this end, we first undertake an analysis of the continuous search space of the PSP problem for generating the landscape structure and proposed a random walk algorithm. Taking the cue from the analysis of the search space, we generated protein landscape structure based on sampling technique for determining the structural properties of the protein landscape. Next, we developed a variant of the differential evolution (DE) algorithm in order to overcome the difficulties associated with the PSP problem without imposing any serious additional computational burden. Drop Developed Variants of particle swarm optimization (PSO), bees algorithm (BA), biogeography-based optimization (BBO), and harmony search (HS) algorithms are developed in order to overcome the difficulties associated with the landscape structure while solving the PSP problem. The proposed algorithms are compared with several state-of-the-art PSP algorithms over many artificial and real data sets reflect the superior performance of the proposed schemes in terms of final accuracy, speed, and robustness. We also study the implication of the proposed hybrid technique which is configured based on the structural features for obtaining the better minimum energy value of the PSP problem.

The book is divided into eight chapters. Chapter 1 presents the overview of the PSP problem and discusses the importance to the research community in the domain of computational biology. The chapter begins with a discussion of the basic concepts of proteins structure and gradually elaborates the various terminology about the structure prediction of a protein. It also discusses the computational models. Next, the chapter provides the basic fundamentals and different types of metaheuristic algorithms. It explains some widely used metaheuristic techniques such as DE, PSO, BA, BBO, and HS in details.

Chapter 2 presents a literature review of metaheuristic techniques based PSP, a scope of the work and contribution to the book chapters. Initially, a critical analysis of the existing metaheuristic techniques employed in PSP problem are presented. Next, we explain the scope of the work. The main contributions of the book are briefly outlined at the end of the chapter.

Chapter 3 investigates continuous fitness landscape structure with random walk (RW) algorithm. The chapter develops two chaos-based random walk (CRW) algorithm for generating sample points in the search space and fitness landscape is created based on the relative fitness of the neighboring sample points. The chaotic map is used to generate the chaotic pseudo-random numbers (CPRN) for determining variable scaled step size and direction of the proposed RW algorithm. Histogram analysis provides better coverage of the search space by the CRW algorithm compared to the existing random walk algorithms in the literature. The chapter also presents empirical results over complex benchmark functions and PSP problem to validate the efficiency of the proposed method.

Chapter 4 determines the characteristics of PSP problem or its structural features using FLA based on which the most appropriate algorithm can be recommended for solving the problem. The chapter describes protein landscape structure generated by using the quasi-random sampling technique and city block distance. The structural properties of the PSP problem are analyzed with the help of information and statistical landscape measures. The chapter also reports an important investigation of six well-known real-coded optimization algorithms over the same set of protein sequences and the performances are subsequently analyzed based on the structural features. Finally, it suggests the most appropriate algorithms for solving the PSP problem.

Chapter 5 describes a DE-based metaheuristic algorithm for solving the PSP problem. In the proposed method, parameters of the DE algorithm are controlled adaptively using Lévy distribution. The distribution function is used to made a possible changes in the control parameters of DE adaptively in order to achieve balance between exploration and exploitation strategy in the search space. The performance of the proposed method has been extensively compared with some parameter control techniques over a test-suite of expanded multimodal, hybrid composite functions and the different protein instances. It is apparent from the computer simulations that proposed method provides significant performance in terms of accuracy and convergence speed to obtain global minimum energy.

Chapter 6 provides four variants of metaheuristic algorithms for solving the PSP problem in both two-dimensional and three-dimensional AB off-lattice model. The proposed algorithms are configured in order to tackle the structural features reported in Chap. 4. The chapter begins with a modified version of the classical PSO algorithm which uses a local search technique. Use of local search makes it possible to jump out from the local optima and prevent the premature convergence. Next, a mutation-based BA is developed for enhancing the diversity in the search space to solve the complex problem like the PSP problem. The chapter also describes chaos-based BBO and perturbation-based HS algorithms which are configured to prevent the loss of diversity in the search space and being trapped in local optima when dealing with PSP problem. Experimental results over the artificial and real protein sequences of varying range of protein lengths indicate that the proposed algorithms are very efficient to solve the PSP problem.

Chapter 7 explains a hybrid technique that combines the merits of algorithms to improve the performance of an optimizer when dealing with multimodal problem. The chapter develops a synergism of the improved version of the PSO and DE algorithm. The proposed method is executed in an interleaved fashion for balancing exploration and exploitation dilemma in the evolution process. Experiments are carried out on the real protein sequences with different lengths taken from Protein Data Bank (PDB) and the results indicate that the proposed method outperforms the state-of-the-art algorithms.

Finally, Chap. 8 concludes with the self-review of the book chapters and scope of the future research.

Durgapur, West Bengal, India   Nanda Dulal Jana
November 2017   Swagatam Das
                                                                                         Jaya Sil

Contents

Abbreviations and Symbols

Abbreviations

ABC	Artificial Bee Colony
APM	Adaptive Polynomial Mutation
BA	Bees Algorithm
BBO	Biogeography-Based Optimization
CLPSO	Comprehensive Learning Particle Swarm Optimization
CMA-ES	Covariance Matrix Adaptation Evolution Strategy
CR	Crossover Ratio
CRW	Chaotic Random Walk
DE	Differential Evolution
DMP-HS	Difference Mean-based Perturbation Harmony Search
EA	Evolutionary Algorithm
FDC	Fitness Distance Correlation
FLA	Fitness Landscape Analysis
GA	Genetic Algorithm
HM	Harmony Memory
HMCR	Harmony Memory Considering Rate
HMS	Harmony Memory Size
HP	Hydrophobic-Polar
HPSODE	Hybrid PSO and DE
HS	Harmony Search
HSI	Habitat Suitability Index
LdDE	Lévy distributed Differential Evolution
NMR	Nuclear Magnetic Resonance
PAR	Pitch Adjusting Rate
PSL	Protein Sequence Length
PSO	Particle Swarm Optimization
PSOLS	Particle Swarm Optimization with Local Search
PSP	Protein Structure Prediction
QPSO	Quantum-based Particle Swarm Optimization

RW	Random Work
SI	Swarm Intelligence
SIV	Suitability Index Variable
TS	Tabu Search
TSM	Tabu Search Mechanism
TSR	Tabu Search Recombination

Symbols

$\mathbf{H}$	Hydrophobic or nonpolar amino acid
$\mathbf{P}$	Hydrophilic or polar amino acid
θ_i	The ith bend angle
β_i	The ith torsion angle
r_{ij}	The distance between i^{th} and j^{th} amino acid of the chain
ξ_i	The ith amino acid
t_{max}	The maximum generations or iterations
$\mathbf{x_{i,t}}$	The ith vector at t^{th} generation
x_j^{min}	The lower bound of the jth component of a vector
x_j^{max}	The upper bound of the jth component of a vector
D	Dimensions of the problem
NP	The population size
λ_i	The ith immigration rate
μ_i	The ith emigration rate
$\mathbf{v}_i(t)$	The velocity of the ith particle at generation t
ω	An inertia weight
F	Fitness landscape
$\mathbf{x}$	Vector
f_i	The ith fitness value
$N_\phi(\mathbf{x}_i)$	Neighborhood of $\mathbf{x}_i$
$\mathbf{x}^*$	The global minimum
$\mathbb{R}$	Real number
$\sum$	Sum
$\prod$	Product

List of Figures

List of Tables

List of Algorithms

Chapter 1
Backgrounds on Protein Structure Prediction and Metaheuristics

Abstract This chapter provides a comprehensive overview of the protein structure prediction problem based on metaheuristic algorithms. At first, the basic concepts of proteins, the level of protein structure have been presented in a formal way. A computational model, as well as techniques, have been addressed for solving protein structure prediction (PSP) problem. The chapter discusses the basic fundamentals of metaheuristics algorithms in detail and finally ends with a discussion of techniques are used in the book towards solving the problem.

1.1 Introduction

Proteins are the main building blocks of dry mass cells in all living organisms. All cell activities are performed with the help of proteins. Proteins are complex macro-molecules combining with amino acids that are connected by peptide bonds. During synthesis, an individual protein is folded into a three-dimensional structure [1], provides essential biological functions and properties which are playing important role in biological science, medicine, drug design and carries disease prediction, pharmaceuticals and much more [2, 3]. Sometimes, proteins may fold into an incorrect structure (known as misfolding) leads to a protein with the different properties which can be harmful to the organisms [4–8]. The knowledge of protein structure is, therefore, crucial in protein research.

Biologists have defined a hierarchy of protein structures with four levels in order to better characterize the structure properties. The *primary structure* is simply the linear protein chain, i.e. a sequence of amino acids. The *secondary structure* is the local conformation of protein structures. There are three types of major secondary structures, known as alpha-helix, beta-sheet, and coils (or loops). The *tertiary structure* is the global three-dimensional structures. Sometimes, multiple protein chains unit together and form hydrogen bonds between each other, resulting in the

© Springer International Publishing AG 2018
N. D. Jana et al., *A Metaheuristic Approach to Protein Structure
Prediction*, Emergence, Complexity and Computation 31,
https://doi.org/10.1007/978-3-319-74775-0_1

quaternary structures. Thus, the protein structure prediction (PSP) is the meaning of determining the three-dimensional structure of a protein from its primary structure i.e. a sequence of amino acids.

In general, the experimental techniques such as X-ray crystallography and Nuclear Magnetic Resonance (NMR) have been used to determine a protein structure from its sequence of amino acids [9, 10]. However, these experimental techniques are not always feasible due to very expensive, time-consuming, strict laboratory requirements and heavy operational burdens [11]. As a result, the gap between the protein sequences and the known structures are increasing rapidly day by day [12]. For example, there are 78028152 protein sequences but the structure is known only for 117184 of them as of January 2017 [13]. As a rough estimation of a protein structure is usually more useful than no structure at all, a number of computational methods have been developed for PSP in the last two decades [14–16].

The computational approaches are based on utilization of computer resources and free from laboratory activities. These methods have the potential to correlate and predict the structure of a protein from its primary sequence in order to overcome the difficulties associated with the experimental approaches. Based on template or template-free modeling, computational approaches are dived into three category [15]: *homology or comparative modeling*, *threading or fold recognition* and *ab-initio or denovo method*. Homology modeling is based on the similarity comparison of the sequence while threading approach is a process to thread together likely short sub-conformation of the corresponding sub-sequence. These two techniques are totally depended on the availability of similar sequence in the database i.e. these are template based methods and the results may become unconvincing for dissimilar sequences [17]. Consequently, the *ab-initio* method predicts three-dimensional structure of an individual protein from its primary sequence directly based on intrinsic properties (namely, hydrophobic and hydrophilic) of amino acids i.e. template-free approach. The concept of *ab-initio* structure prediction method is based on the *Anfinsen's thermodynamic hypothesis* where the theory of thermodynamics [18] states that conformation of a protein corresponds to the global minimum free energy surface which is called a native structure of the protein. Thus, a native structure is a stable structure of a protein which always consume minimum energy among all the conformations of the protein. Therefore, PSP based on *ab-initio* approach can be transformed into a global optimization problem where energy function would be minimized.

The *ab-initio* based protein structure prediction has two important key parts. The first part is to state the simplified mathematical model that corresponds to a protein energy function. The second part is to develop a computationally efficient optimization algorithm which can find the global minimum of the potential energy function. A number of simplified models that derived the protein energy function for the PSP have been proposed in [19–22]. The simplest computational model for the PSP problem is known as Hydrophobic-Polar (HP) or lattice model, both are two and three dimensions. Another widely used physical model that extended from former model known as the *AB* off-lattice model. The off-lattice model reveals the structure of proteins at much higher atomic details and represents in both two and three dimensions [21]. Unfortunately, the models are used in computational techniques that belong to a class of difficult optimization problem known as *NP*-hard problems [23–26] that means there is no polynomial-time algorithm to solve the PSP problem in either 2D or 3D dimension. Therefore, the *ab-initio* structure prediction is very challenging task in computational biology.

The metaheuristic optimization algorithms have potential ability to solve the *NP*-hard problem efficiently. Recent advances are a result of the application of meta-heuristic approaches which are used to find the native conformation of a protein based on a physical model. Since the primary sequence of amino acids determines the ultimate three-dimensional structure of proteins, it is essential to design and develop an effective optimization method to predict the protein structures, which is the main task of the book.

The rest of this chapter is arranged as follows. In Sects. 1.2, 1.3, 1.4, 1.5 and 1.6, the basic concepts related to protein structure prediction are explained. Section 1.7 provides an overview of several prominent metaheuristic algorithms for global numerical optimization. Finally, the organization of the book is presented in Sect. 1.8.

1.2 Proteins and Amino Acids

Proteins are complex organic macromolecules of which basic structural unit is the amino acid. An amino acid is a chemical compound which consists of an amino group (NH_2), a carboxyl group (COOH) and a unique R group as shown in Fig. 1.1. All the groups are bonded to the central carbon atom known as α-carbon (C_α) atom. The 'R' chemical group constitutes side chain which is also known as 'R'-residue specifying the type of the amino acid. There are 20 types of amino acids commonly found in proteins which differ in size, shape, charge hydrogen-bonding capacity, hydrophobic and reactivity, as characterized by their unique 'R' group.

Two amino acids are connected to form a polypeptide or peptide bond as shown in Fig. 1.2. In the polypeptide bonds, the carboxyl group of one amino acid join with the amino group of another amino acid and produced water molecule (H_2O). The resulting chain is referred to as a peptide chain and used to form a long sequence of amino acids. The start and end terminals of a peptide chain are chemically different.

Fig. 1.1 Basic structure of an amino acid

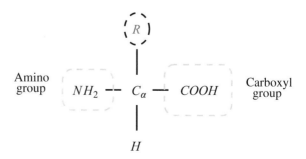

Amino group

NH_2 — C_α — $COOH$

Carboxyl group

R

H

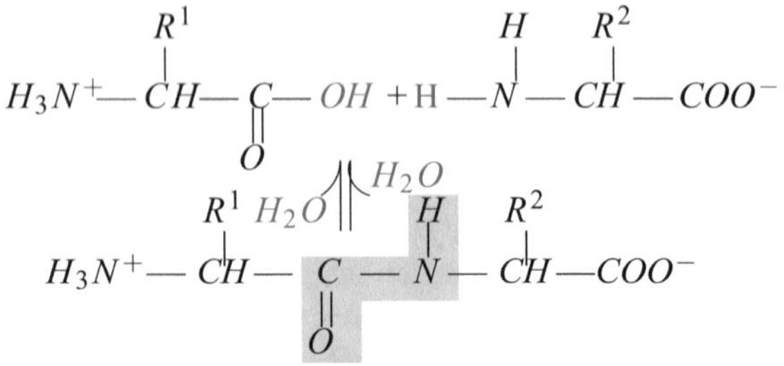

Fig. 1.2 Peptide bonds between two amino acids

$$H_3N^+ - \overset{R^1}{\underset{}{C}}H - \overset{}{\underset{O}{C}} - OH + H - \overset{H}{\underset{}{N}} - \overset{R^2}{\underset{}{C}}H - COO^-$$

$$\overset{H_2O}{\Updownarrow} H_2O$$

$$H_3N^+ - \overset{R^1}{\underset{}{C}}H \; \overset{H_2O}{-} \; \overset{}{\underset{O}{C}} - \overset{H}{\underset{}{N}} - \overset{R^2}{\underset{}{C}}H - COO^-$$

Fig. 1.3 Dihedral and bond angles in protein

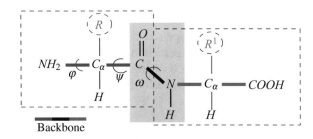

Backbone

Start terminal carrying amino group, called N-terminal whereas end terminal carrying carboxyl group called C-terminal. Thus, a protein always present in the form of 'N' to 'C' direction.

The peptide bond forms a plane which is fairly rigid and planner because there is no free rotation around the peptide bond. There is more flexibility has been found in the protein chain and represented by an angle. The rotation around the bond, N–C_α bond known as ϕ angle, C_α–C bond known as ψ angle and another angle ω exists between C–N (peptide) bonds. All angles are depicted in Fig. 1.3, while the values of angles are restricted to small regions [27].

1.3 Level of Protein Structures

There are four different levels of protein structure hierarchic, clearly identified in most protein structures as shown in Fig. 1.4. Level of hierarchic can be described from the linear chain of amino acids sequence to complex 3D structure [28]. An important property of the protein folding process is that protein sequences have been selected by the evolutionary process to achieve a reproducible and stable structure.

1.3.1 Primary Structure

Linear sequence of amino acids connected by peptide bonds is referred to *primary structure* of protein, which doesn't convey any geometrical information in the structure.

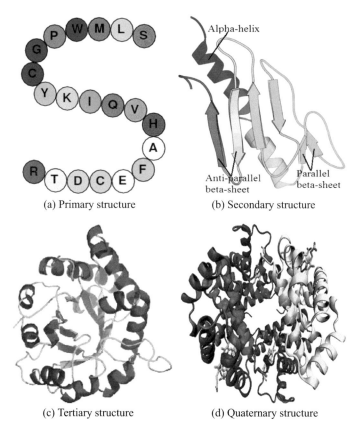

(a) Primary structure   (b) Secondary structure

(c) Tertiary structure   (d) Quaternary structure

Fig. 1.4 Different levels of structures of a protein (Figures are adopted from http://goo.gl/images/KPbnr0)

1.3.2 Secondary Structure

The *secondary structure* of a protein can be thought of as the local conformation of the polypeptide chain, or intuitively as building blocks for its three-dimensional structures. There are two types of secondary structures dominant in this local conformation: α-helix, a rod-shaped peptide chain coiled to form a helix structure, and β-sheets, two peptide strands aligned in the same direction (parallel β-sheet) or opposite direction (anti-parallel β-sheet) and being stable by hydrogen bonds (Fig. 1.4b). These structures exhibit a high degree of regularity and they are connected with the rest irregular regions, referred to as coil or loop.

1.3.3 Tertiary Structure

The *tertiary structure* of a protein is often defined as the global three-dimensional structure and usually represented as a set of 3-D coordinates for each atom. It is widely believed that the side-chain interactions ultimately determine how the secondary structures are combined to produce the final structure.

1.3.4 Quaternary Structure

The *quaternary structure* is the stable association of multiple polypeptide chains resulting in an active unit. Not all proteins can exhibit quaternary structures. However, it is found that the quaternary structures are stabilized mainly by the same non-covalent interactions as tertiary structures, such as hydrogen bonding, van der walls interactions, and ionic bonding.

1.4 Protein Structure Prediction

A protein can fold into specific three-dimensional structure represented as conformation using the freedom of rotations of the peptide chains. The final structure is known as a native structure which consumes minimal free energy [18]. The PSP is the problem of determining the native structure of a protein, given its sequence of amino acids. Structure prediction of a protein is one of the most important tasks in computational biology with the goal of being able to understand its functions and the mechanism of action.

1.5 Protein Structure Prediction Methods

1.5.1 Experimental Methods

There are two experimental methods mostly used for protein structure determination, X-ray crystallography and Nuclear Magnetic Resonance (NMR). In X-ray crystallography [29], the crystallization processes usually take months to years and there exist no rules about the optimal conditions for a protein solution that results in a good protein crystal. On the other hand, NMR [30] does not require protein crystal but treats the protein in solution. Experimental methods are also time consume, laborious and cost expensive. Moreover, these methods are not always feasible to conduct due to time constraint, strict laboratory requirements and heavy operational burdens [11, 31], resulting in a huge gap between sequence and structure knowledge, called *sequence-structure gap* [32]. To bridge this sequence-structure gap, protein structure prediction preferred using computational approaches which would take less cost and time [33, 34].

1.5.2   Computational Methods

The computational methods are characterized by utilization of computer resources and free from laboratory activities and have potential abilities to overcome the difficulties associated with the experimental approaches. Significant research interest [33] is growing into an application of computational methods for protein structure prediction. The main computational methods for PSP can be divided into three category [15] such as *homology or comparative modeling* [35], *threading or fold recognition* [36] and *ab-initio* or *denovo* approach [17]. This section provides an overview of these approaches.

1.5.2.1   Homology Modeling

This approach is based on the similarity comparison of the sequence, also known as comparative modeling. If a sequence of unknown structure (known as target sequence) can be aligned with one or more sequences of known structures (known as template structures) then the known structures are used to predict the structure adopted by the target sequence. The very first step of the approach is to find template protein related to the target. After selecting the templates, a sequence alignment algorithm is used to align the target sequence with the template structures. The accuracy of this method is determined by the accuracy of the alignment. If the accuracy is 70% and above, the target and the template sequences are closely related and demonstrate good alignments.

1.5.2.2   Threading Approach

Threading is the protein structure prediction method applied when the sequence has little or no primary sequence similarity to any sequence in a known structure and/or model of the structure library represents the true fold of the sequence. It is a process of finding similar folds and not similar sequences. At first, the method tries to find out known protein structure and search the most compatible fold with the query sequence whose structure is to be predicted from a library. An optional match between the query sequence and a structural model is to be determined by alignment algorithms and evaluate the quality of a fold using an objective function. The top scoring alignments provide possible templates. Many methods have been developed based on *hidden markov model* [37] and Chou-Fasman model [38] for threading approach. The threading approach outperforms the comparative modeling for sequence similarities below 25% [39].

The aforementioned methods totally depend on the availability of similar sequence in the database i.e. these are template-based methods. Moreover, their results may become unconvincing for dissimilar sequences [17] and they become less accurate for longer sequences. The formation of the whole conformation as derived from its sub-conformations is less likely to match the native conformation because more dissimilarity is added between the similar fragments [40, 41].

1.5.2.3   Ab-initio Approach

The *ab-initio* or *denovo* method is a template-free modeling approach. It predicts the three-dimensional native conformation from its primary sequence alone based on hydrophobic and hydrophilic properties of amino acids [42]. The concept of *ab-initio* method is based on Anfinsens thermodynamics hypothesis which states that the native structure corresponds to the global minimum of a free energy function. Evaluation of all possible conformation is needed and the structure with minimum energy is treated as a native structure. So, representation of conformation for a protein is very important in *ab-initio* approach. Each conformation is accessed by an energy function and finally, a searching technique is needed to determine native confirmation from the set of all possible confirmations. However, in practice *ab-initio* approach is computationally expensive because existing energy functions have limited accuracy as conformational searched space increases with protein sequence length. In this book, we focus on *ab-initio* protein structure prediction.

1.6   Computational Model

A computational approach for protein structure prediction demands a model that represents it abstractly, in a given level of atomic interactions. Based on well-established thermodynamical laws, the prediction of the structure of a protein is modeled as the minimization of the corresponding free energy with respect to the possible conformations that a protein is able to attain. Formally, the native conformation of a protein is defined as the state at which the free-energy is minimal. A number of atomic interactions of the structure are to be modeled depends on the choices of the model. The predominant class of simplified models is the so-called lattice models, which are discussed next.

1.6.1   HP Lattice Model

One of the most well-known simplified and studied discrete lattice model is the hydrophobic-hydrophilic (HP) model was proposed by Dill [43] for PSP. In the model, H represents hydrophobic (non-polar or water-hating) while P represents hydrophilic (polar or water-liking) and amino acids are classified into this two category. Amino acids of a protein sequence are placed on a lattice in such a way that each amino acid occupies at one grid point and two consecutive amino acids (called sequential amino acids) are also placed adjacent grid points on the lattice. The HP lattice model has a restriction that any two or more amino acids cannot occupy at the same grid point of the lattice. Thus, protein sequence folding on the model should be in a self-avoiding walk for preventing collisions in structure.

The free-energy of a conformation is inversely proportional to the number of non-local bonds. A non-local bond only takes place when a pair of non-adjacent hydrophobic amino acids of the sequence lie in adjacent positions on the lattice. Consequently, minimizing the free-energy is equivalent to maximize the number of hydrophobic non-local bonds. A simple free-energy function of a conformation, suggested by [44] is defined in Eq. (1.1).

$$E = \sum_{i<j} e(r_i r_j) \Delta(r_i - r_j), \tag{1.1}$$

where r_i and r_j are the i and jth amino acids of the sequence; $\Delta(r_i - r_j) = 1$, if amino acids r_i and r_j have a non-local bond, otherwise $\Delta(r_i - r_j) = 0$. Depending on the type of contact between amino acids, the energy e_{HH}, e_{HP} or e_{PP}, correspond to H–H, H–P or P–P contacts.

In the standard HP lattice model, values for the parameters are: $e_{HH} = -1, e_{HP} = 0$ and $e_{PP} = 0$. Figure 1.5, presents a conformation of polypeptide with 20 amino acids using the 2D-HP model. Black and white circles in Fig. 1.5 represent the hydrophobic (H) and hydrophilic (P) amino acids, respectively. The start circle is the first amino acid of the sequence. The chain is connected by solid (black color) lines, and the non-local bonds are represented by read color lines in Fig. 1.5, shows the eight H–H contacts.

1.6.2 AB Off-Lattice Model

The AB off-lattice model, known as toy model was proposed by Stillinger et al. [21] in 1993 for predicting the structure of a protein from its primary sequence of amino acids. AB off-lattice model has been widely used for its simplicity and inter-molecular interactions among the amino acids. In AB off-lattice model, 20 amino acids are classified into hydrophobic and hydrophilic residues, labeled as 'A' and 'B' respectively. Two residues are linked by rigid unit-length bonds and angle between two bonds changes freely in a two-dimensional Euclidean space. An n length pro-tein sequence is represented by $(n - 2)$ bend angles $\theta_2, \theta_3, \ldots, \theta_{n-1}$ at each of the non-terminal residues. The bend angle $\theta_i \in [-180°, 0)$ and $\theta_i \in (0, 180°]$ represents rotation of amino acids in clockwise and counter clockwise direction, respectively while $\theta_i = 0$ represents linearity in the successive bonds. Figure 1.6 shows the position of amino acids for a protein sequence $AABBABAB$ of length eight which is specified by 6 $(8 - 2)$ bend angles $(\theta_2, \theta_3, \ldots, \theta_7)$. Amino acids along the backbone can be encoded by a set of bipolar variables ξ_i. If the ith amino acid is 'A' then $\xi_i = 1$ and

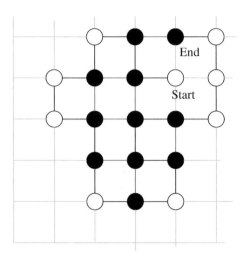

Fig. 1.5 An example of conformation of 20 amino acids using the 2D-HP lattice model

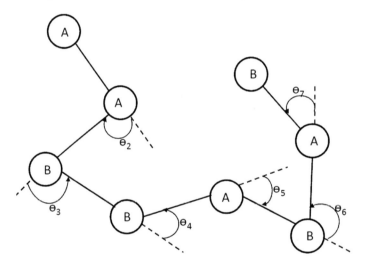

Fig. 1.6 An example of a 2D AB off-lattice model for a protein sequence *AABBABAB*

$\xi_i = -1$ for 'B'. Therefore, the intra-molecular potential energy function Φ for any n length protein sequence can be expressed using Eq. (1.2).

$$\Phi = \sum_{i=2}^{n-1} V_1(\theta_i) + \sum_{i=1}^{n-2} \sum_{j=i+2}^{n} V_2(r_{ij}, \xi_i, \xi_j), \tag{1.2}$$

where θ_i is the bend angle and $V_1(\cdot)$ is the bending potential energy as defined in Eq. (1.3).

$$V_1(\theta_i) = \frac{1}{4}(1 - cos(\theta_i)). \tag{1.3}$$

The non-bonded interactions $V_2(\cdot)$ have a species - dependent Lennard - Jones (12, 6) form, described in Eq. (1.4).

$$V_2(r_{ij}, \xi_i, \xi_j) = 4[r_{ij}^{-12} - C(\xi_i, \xi_j)r_{ij}^{-6}], \tag{1.4}$$

where $C(\xi_i, \xi_j) = \frac{1}{8}(1 + \xi_i + \xi_j + 5\xi_i\xi_j)$ and r_{ij} denotes the distance between i and jth residue of the chain. For an AA pair, $C(\xi_i, \xi_j) = 1$ regarded as strong attraction, for an AB or BA pair, $C(\xi_i, \xi_j) = -0.5$, regarded as weak repelling and for a BB pair, $C(\xi_i, \xi_j) = 0.5$, regarded as weak attraction between the residues.

The 3D AB off-lattice model is extended from 2D AB off-lattice model by adding torsion angle to each bond [45]. The three dimensional structure of an n length protein is specified by the $(n - 1)$ bond length, $(n - 2)$ bend angles $\theta_1, \theta_2, \ldots, \theta_{n-2}$ and $(n - 3)$ torsion angles $\beta_1, \beta_2, \ldots, \beta_{n-3}$. Therefore, the n length protein conformation is determined by $(2n - 5)$ angle parameters $(\theta_1, \theta_2, \ldots, \theta_{n-2}, \beta_1, \beta_2, \ldots, \beta_{n-3})$. Cartesian coordinates of the ith residue in the 3D AB off-lattice model is obtained using Eq. (1.5).

$$(x_i, y_i, z_i) = \begin{cases} (0, 0, 0), & \text{for } i = 1 \\ (0, 1, 0), & \text{for } i = 2 \\ (cos(\theta_1), sin(\theta_1) + 1, 0), & \text{for } i = 3, \text{and} \\ (x_{(i-1)} + cos(\theta_{i-2}) \times cos(\beta_{i-3}), \\ y_{(i-1)} + sin(\theta_{i-1}) \times cos(\beta_{i-3}), \\ z_{(i-1)} + sin(\beta_{i-3})), & \text{for } 4 \leq i \leq n \end{cases} \tag{1.5}$$

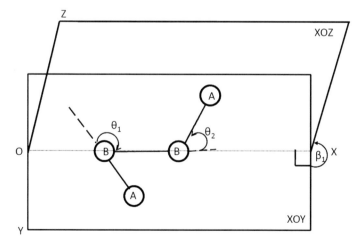

Fig. 1.7 An example of a 3D AB off-lattice model for a protein sequence *ABBA*

here, the first three residues are defined in the plane, $z = 0$. Then position of the remaining residues ($i \geq 4$) are calculated on the basis of the position of the previous one. Figure 1.7 shows the locations of the amino acids for a protein sequence *ABBA* which is determined by the angle vector $(\theta_1, \theta_2, \beta_1)$.

For a protein sequence, the minimum protein energy (Φ) is obtained through optimal settings of the angles $(\theta_1, \theta_2, \ldots, \theta_{n-2}, \beta_1, \beta_2, \ldots, \beta_{n-3})$, associated in protein energy function.

1.7 Population Based Metaheuristic Techniques

The heuristic is defined in the context for solving a problem without the exhaustive application of a procedure. It is a method that looks for an approximate solution, need not particularly have a mathematical convergence proof, and applicable to the problems which are not solved by traditional optimization methods. A metaheuristic method in the context of solving search and optimization problems uses one or more heuristics and satisfying all the properties of the heuristic algorithm. Thus, a metaheuristic method [46] (i) seeks to find a near-optimal solution, instead of specifically trying to find the exact optimal solution, (ii) usually has no rigorous proof of convergence to the optimal solution, and (iii) computationally faster than exhaustive search. Metaheuristic methods are iterative in nature and often use stochastic operations in searching in order to modify one or more initial candidate solutions which are generated randomly. Especially, for complicated problems or large problem instances, metaheuristics are often able to offer a better trade-off between solution quality and computing time. Moreover, metaheuristics are more flexible than exact methods in two important ways. First, metaheuristic algorithms can be adapted to fit the needs of most real-life optimization problems in terms of expected solution quality and allowed computing time. Secondly, metaheuristics do not put any demands on the formulation of the optimization problem like requiring constraints or objective functions to be expressed as linear functions of the decision variables. The ability of the metaheuristic methods to handle complexities associated with practical problems and arriving at a reasonably acceptable solution is the main reason for its popularity in the recent past [46].

Most metaheuristic methods are motivated by natural, physical or biological principles and try to mimic them at a fundamental level through various operators. Two very important concepts that largely determine the behavior of a metaheuristic method are *exploration* and *exploitation*. Exploration refers to how well the operators diversify solutions in the search space while exploitation refers to how well the operators are able to utilize the information available from solutions to intensify the search. Such exploration and exploitation give the metaheuristic a global search and local search characteristic. Certain advantages which characterize metaheuristics over the classical optimization methods as follows [46]:

- Metaheuristics can lead to good enough solutions for computationally easy problems with large input complexity, which can be a hurdle for classical methods;
- Metaheuristics can lead to good enough solutions for the NP-hard problems;
- Unlike most classical methods, metaheuristics require no gradient information and therefore can be used with non-analytic, black-box or simulation-based objective functions;
- Most metaheuristics have the ability to recover from local optima due to inherent stochasticity.

1.7.1 An Optimization Problem

Optimization problems can be divided into two categories depending on whether the variables are *discrete* or *continuous*. An optimization problem with discrete variables is known as a *combinatorial* optimization problem. Problems with continuous variables include constrained problems and multi-modal problems. In the simplest case, an optimization problem consists of maximizing or minimizing a function by systematically choosing input values from within an allowed set and computing the value of the function. A standard optimization (continuous) problem can be formally stated as follows:

$$
\begin{aligned}
\underset{x \in \mathbf{x}}{\text{minimize}} \quad & f(\mathbf{x}) \\
\text{subject to} \quad & g_i(\mathbf{x}) \leq 0, \quad i = 1, \ldots, l_1, \\
& h_i(\mathbf{x}) = 0, \quad i = 1, \ldots, l_2, \\
\text{and} \quad & x_i^{min} \leq x_i \leq x_i^{max}, \quad i = 1, \ldots, n,
\end{aligned}
\tag{1.6}
$$

where $f(\mathbf{x}) : U \rightarrow \mathbb{R}, U \subseteq \mathbb{R}^n$ is an objective function (or fitness function) to be minimized over the vector of variables $\mathbf{x}$, $g_i(\mathbf{x}) \leq 0$ represent l_1 number of inequality constraints and $h_i(\mathbf{x}) = 0$ represent l_2 number of equality constraints. The domain U of f is called the search space, while the elements of U are called *candidate solutions*. A candidate solution is feasible if all $(l_1 + l_2)$ constraints and variable bounds $[x_i^{min}, x_i^{max}]$ are satisfied.

In mathematics, conventional optimization problems are usually stated in terms of minimization. A local minimum $\mathbf{x}^*$ is defined as a point for which there exists some $\delta > 0$ so that for all x such that $\|\mathbf{x} - \mathbf{x}^*\| \leq \delta$, the expression $f(\mathbf{x}^*) \leq f(\mathbf{x})$ holds; that is on some region around $\mathbf{x}^*$ all of the function values are greater than or equal to the value at that point.

1.7.2 The Metaheuristic Family

Scientists and engineers from all disciplines often have to deal with the classical problem of search and optimization. An optimization problem searches the best-suited solution from some set of available alternatives within the given constraints and flexibilities by formulating an objective function. A feasible solution that minimizes (or maximizes, if that is the goal) the objective function is called a *optimal solution*. Generally, unless both the objective function and the feasible region are convex in a minimization problem, there may be several local minima. A local minimum is at least as good as any nearby points while a global minimum is at least as good as every feasible point. In a convex problem, if there is a local minimum that is interior (not on the edge of the set of feasible points), it is also the global minimum, but a non-convex problem may have more than one local minimum, not all of which need be global minima. Unfortunately, most real-world problems are non-convex and NP-hard, which makes metaheuristic techniques a popular choice. The nature of the optimization problem plays an important role in determining the optimization methodology to be used [46].

There are different ways to classify and describe metaheuristic algorithms, depending on the characteristics. The most common characteristic used for classifying the metaheuristics is based on the number of solutions used at the same time and modified in subsequent generations. Single-solution metaheuristics start with one initial solution which gets modified iteratively [46]. However, the modification process itself may involve more than one solution, but only a single solution is used in each following generation. Population-based metaheuristics use more than one initial solution to start optimization. In each generation, multiple solutions get modified, and some of them make it to the next generation. Modification of the solutions is performed through operators that often use special statistical properties of the population.

Another way of classifying metaheuristics which are based on domain specific that mimicking bio-inspired and nature-inspired phenomena, further sub-categorized as evolutionary algorithms, swarm intelligence based algorithms and physical phenomenon based algorithms [46]. Evolutionary algorithms (like genetic algorithms, evolution strategies, differential evolution, genetic programming, evolutionary programming, etc.) mimic various aspects of evolution in nature such as survival of the fittest, reproduction and genetic mutation. Swarm intelligence algorithms mimic the group behavior and/or interactions of living organisms (like ants, bees, birds, fireflies, glowworms, fishes, white blood cells, bacteria, etc.) and non-living things (like water drops, river systems, masses under gravity, etc.). The rest of the metaheuristics mimic various physical phenomena like annealing of metals, musical aesthetics (harmony), etc. In the following section, we mainly discuss some of the most promising evolutionary algorithms, swarm intelligence techniques and a metaheuristic technique based on a physical phenomenon.

1.7.3 The Evolutionary Algorithms

Evolutionary algorithms (EAs) are inspired by Darwinian principle. Natural selection and adaptation in Darwinian principle are the key sources of inspiration driving the

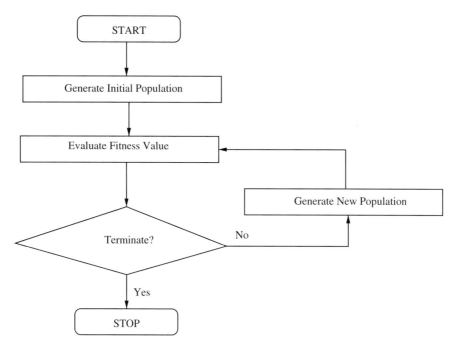

Fig. 1.8 Flowchart of general evolutionary algorithm

EA candidate solutions towards the optimum by 'survival of the fittest'. To be more precise an EA consists of a population of individuals each having a fitness value and a genome representing one candidate solution to the given problem. The general principle of an EA with a simple flowchart is given in Fig. 1.8.

1.7.3.1 Differential Evolution (DE)

The DE algorithm is a population based stochastic search and optimization technique for solving global optimization problems [47–49] in iterative way. A population of size *NP* with *D*-dimensional parameter vectors are supplied for initializing the DE procedure and each *D*-dimensional parameter vector is known as individual of the population. The generation (t) executes maximum number of generations (t_{max}) to obtain global optimal solution. The ith individual of the population at tth generation is denoted as: $\mathbf{x}_{i,t} = \{x_{1,i,t}, x_{2,i,t}, \ldots, x_{D,i,t}\}$. Here $x_{j,i,t}$ is the jth component ($j = 1, 2, \ldots, D$) of the ith parameter vector ($i = 1, 2, \ldots, NP$) at generation t and constrained by the predefined minimum and maximum bounds say, x_j^{min} and x_j^{max} respectively. For example, the initial value of jth parameter in ith individual at generation $t = 1$ is calculated using Eq. (1.7),

$$x_{i,j} = x_j^{min} + (x_j^{max} - x_j^{min})\, rand\,(0, 1), \tag{1.7}$$

where $rand(0, 1)$ represents a uniformly distributed random variable within the range [0, 1]. After initialization of the parameter vectors, DE enters in a cycle to execute the following steps sequentially.

Mutation with Difference Vectors:

After initialization, DE creates a mutant vector or donor vector, $\mathbf{V}_{i,t}$ corresponding to each individual or target vector $\mathbf{x}_{i,t}$ in the current population through mutation. The ith donor vector of the current generation, t is represented as: $\mathbf{V}_{i,t} = \{v_{1,i,t}, v_{2,i,t}, \ldots, v_{D,i,t}\}$. Most frequently used mutation strategies are listed below [48, 50]:

$$\text{"DE/rand/1"} : \mathbf{V}_{i,t} = \mathbf{x}_{r_1^i,t} + F(\mathbf{x}_{r_2^i,t} - \mathbf{x}_{r_3^i,t})$$

$$\text{"DE/best/1"} : \mathbf{V}_{i,t} = \mathbf{x}_{best,t} + F(\mathbf{x}_{r_1^i,t} - \mathbf{x}_{r_2^i,t})$$

$$\text{"DE/current to best/1"} : \mathbf{V}_{i,t} = \mathbf{x}_{i,t} + F(\mathbf{x}_{best,t} - \mathbf{x}_{i,t}) + F(\mathbf{x}_{r_1^i,t} - \mathbf{x}_{r_2^i,t}), \quad (1.8)$$

$$\text{"DE/best/2"} : \mathbf{V}_{i,t} = \mathbf{x}_{best,t} + F(\mathbf{x}_{r_1^i,t} - \mathbf{x}_{r_2^i,t}) + F(\mathbf{x}_{r_3^i,t} - \mathbf{x}_{r_4^i,t})$$

$$\text{"DE/rand/2"} : \mathbf{V}_{i,t} = \mathbf{x}_{r_1^i,t} + F(\mathbf{x}_{r_2^i,t} - \mathbf{x}_{r_3^i,t}) + F(\mathbf{x}_{r_4^i,t} - \mathbf{x}_{r_5^i,t})$$

where the indices $r_1^i, r_2^i, r_3^i, r_4^i$ and r_5^i are mutually exclusive numbers selected randomly from the range $[1, NP]$ and different from the ith vector. Here, scale factor $F \in [0, 2]$ is a positive control parameter, known as scaling factor used to scale the difference vectors while $\mathbf{x}_{best,t}$ is the best individual vector in the population at generation t with respect to the fitness value.

Crossover:

Crossover operation is performed on each pair of the target vector $\mathbf{x}_{i,t}$ and the donor vector $\mathbf{V}_{i,t}$ to generate a trial vector $\mathbf{u}_{i,t} = \{u_{1,i,t}, u_{2,i,t}, \ldots, u_{D,i,t}\}$. Crossover operator plays an important role to explore the search space of the DE algorithm. There are mainly two types of crossover techniques: binomial and exponential crossover [51]. The binomial crossover is performed on each dimension (D) of each individual depending on the crossover ratio (CR). The trial vector is obtained through the binomial crossover using Eq. (1.9).

$$u_{j,i,t} = \begin{cases} v_{j,i,t} & \text{if } (rand(0, 1) \leq CR \parallel j = j_{rand}) \\ x_{j,i,t} & \text{otherwise.} \end{cases} \quad (1.9)$$

Where $rand(0, 1) \in [0, 1]$, a uniformly distributed random number and $CR \in [0, 1]$. Crossover operator copies the jth parameter of the donor vector $\mathbf{V}_{i,t}$ to the corresponding element in the trial vector $\mathbf{u}_{j,i,t}$ if ($rand \leq CR \parallel j = j_{rand}$), otherwise, it is simply copied from the corresponding target vector $\mathbf{x}_{i,t}$.

In exponential crossover, first an integer $n \in [1, D]$ is chosen randomly, as a starting point in the target vector from where the crossover or exchange of components starts with the donor vector. Another integer L is chosen from the interval $[1, D]$ where

L denotes the number of components in the donor vector which actually contributes to the target vector. Finally, the trial vector is calculated using Eq. (1.10).

$$u_{j,i,t} = \begin{cases} v_{j,i,t} & \text{for } j = \langle n \rangle_D, \langle n+1 \rangle_D, \ldots, \langle n+L-1 \rangle_D \\ x_{j,i,t} & \text{otherwise} \end{cases} , \quad (1.10)$$

where $\langle \rangle_D$ denotes modulo function with modulus D. The integer $L \in [1, D]$ is taken according to the following pseudo-code:
$L = 0$;
Do
$\{L=L+1;\}$
While $((rand(0, 1) < CR)$ and $(L < D))$
Hence in effect, probability $(L \geq v) = (CR)^{v-1}$ for any $v > 0$.

Selection:

After trial vector generation, several parameters may exceed corresponding upper and lower bounds. Parameters can be reinitialized randomly and uniformly with the predefined range prior to application of the selection operator. Selection operator determines whether the target vector $\mathbf{x}_{i,t}$ or the trial vector $\mathbf{u}_{j,i,t}$ survives to the next generation according to their fitness values $f(.)$. For example, if we have a minimization problem, the better individual can be selected as:

$$\mathbf{x}_{i,t+1} = \begin{cases} \mathbf{u}_{i,t} & \text{if } f(\mathbf{u}_{i,t}) \leq f(\mathbf{x}_{i,t}) \\ \mathbf{x}_{i,t} & \text{otherwise.} \end{cases} \quad (1.11)$$

Mutation, Crossover and Selection steps are repeated into subsequent DE generations and continued till a termination criteria is satisfied. Algorithm 1 presents a lay out of the pseudo-code for the classic *DE/rand/1/bin* variant.

1.7.3.2 Biogeography Based Optimization (BBO)

BBO is an evolutionary algorithm (EA) based on the study of the distribution of biological species over time introduced by Dan Simon [52] in 2008. BBO is a new population-based metaheuristic algorithm for solving optimization problems. In BBO, each individual is known as habitat with a habitat suitability index (HSI). The HSI value is obtained by using fitness value to show the measure of goodness of the solution. The high value of HSI represents a better solution where habitat contains a large number of species. On the other hand, the lowest value of HSI represents worst solution which implies the least number of species residing in that habitat. High HSI solutions are more likely to share the features with lower solutions and low HSI solutions are more likely to accept shared the features from good solutions. Information sharing among habitat is known as migration process. In basic BBO, migration process and mutation operation are two main steps, described below.

Algorithm 1 Differential Evolution (DE)

1: Initialize population size (NP), scaling factor (F) and crossover ratio (CR)
2: Randomly initialize a population of NP members.
3: **while** Termination criteria is not satisfied **do**
4:     Evaluate fitness values of all members of the population
5:     **for** $i = 1, 2, ..., NP$ **do**
6:         Generate random integer $rand_i$ between 1 and search space dimension D
7:         **for** $j = 1, 2, ..., D$ **do**
8:             Generate random number $j_{rand} \in [0, 1]$
9:             **if** $rand(0, 1) \leq CR$ or $j = rand_i$ **then**
10:                 $u_{ji} = v_{ji} = x_{jr_1} + F.(x_{jr_2} - x_{jr_1})$
11:             **else**
12:                 $u_{ji} = x_{ji}$
13:             **end if**
14:         **end for**
15:         Evaluate trial vector $\mathbf{u}_i$
16:         **if** $f(\mathbf{u}_i) \leq f(\mathbf{x}_i)$ **then**
17:             Replace $\mathbf{x}_i$ with $\mathbf{u}_i$
18:         **end if**
19:     **end for**
20: **end while**

Migration is a probabilistic operation that improves ith habitat, $\mathbf{H}_i$. It has immigration rate (λ_i) and emigration rate (μ_i) to share features between the habitats. A good solution has higher emigration rate and lower immigration rate and vice versa. The immigration rate (λ_i) and emigration rate (μ_i) of each habitat can be calculated using Eq. (1.12).

$$\lambda_i = I\left(1 - \frac{i}{NP}\right) \quad \text{and} \quad \mu_i = E\left(\frac{i}{NP}\right), \tag{1.12}$$

where I and E are the maximum possible immigration and emigration rate where NP is the maximum number of individuals. Figure 1.9 shows the relation between number of species, immigration rate and emigration rate.

$\mathbf{H}_i$ is selected based on immigration rate while the emigrated habitat $\mathbf{H}_j$ is chosen probabilistically based on μ_j. Habitat exchange their information using migration operation as follows:

$$\mathbf{H}_i(SIV) \leftarrow \mathbf{H}_j(SIV). \tag{1.13}$$

In biogeography, a suitability index variable (SIV) determines the habitability of an island. In BBO, SIV is a feature or a parameter of a solution.

Mutation is a probabilistic operator used to increase the population diversity. In BBO, mutation probability (m_p) is used to replace an SIV in a habitat by a randomly generated SIV. The pseudo code of basic BBO algorithm is shown in Algorithm 2.

Fig. 1.9 Migration model of species

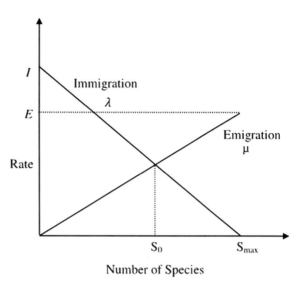

Number of Species

Algorithm 2 Biogeography Based Optimization (BBO)

1: Initialize number of individuals (NP), Dimensions (D)
2: Initialize Population randomly and evaluate fitness values of each individual
3: **while** The stopping criterion is not satisfied **do**
4:     Short the population from best to worst
5:     Calculate the rate of λ_i and μ_i using Eq. (1.12)
6:     **for** each Habitat $i = 1, 2, ..., NP$ **do**
7:         Select habitat $\mathbf{H}_i$ with respect to rate λ_i
8:         **if** rand(0,1) $< \lambda_i$ **then**
9:             **for** each Habitat $j = 1, 2, ..., NP$ **do**
10:                Select habitat $\mathbf{H}_j$ with respect to μ_j
11:                Randomly select a SIV from $\mathbf{H}_j$
12:                Exchange habitat information using Eq. (1.13)
13:            **end for**
14:        **end if**
15:    **end for**
16:    Update the probability for each habitat
17:    **for** $i = 1, 2, ..., NP$ **do**
18:        **for** $j = 1, 2, ..., D$ **do**
19:            **if** rand(0,1) $< m_p$ **then**
20:                Replace j^{th} SIV of the $\mathbf{H}_i$ with a randomly generated SIV
21:            **end if**
22:        **end for**
23:    **end for**
24:    Evaluate HSI for each individual
25: **end while**

1.7.4 Swarm Intelligence Algorithms

The behavior of a single ant, bee, termite and wasp often is too simple, but their collective and social behavior is of paramount significance. The collective and social behavior of living creatures motivated researchers to undertake the study of today what is known as *Swarm Intelligence* (SI). Historically, the phrase swarm intelligence was introduced by Beni and Wang in 1989s [53] in the context of cellular robotics. A group of researchers in different parts of the world started working almost at the same time to study the versatile behavior of different living creatures and especially in the social insects. The efforts to mimic such behaviors through computer simulation finally resulted into the fascinating field of SI. The SI systems are made up of a population of simple agents (an entity capable of performing/executing certain operations) interacting locally with one another and with an environment. Although there is normally no centralized control structure dictating how individual agent should behave, local interactions between such agents often lead to the emergence of global behavior. Two promising SI algorithms are described below which has important application in protein structure prediction problem.

1.7.4.1 Particle Swarm Optimization (PSO)

Particle swarm optimization (PSO) is an efficient search and optimization technique, introduced by Kennedy and Eberhart in 1995 [54, 55]. PSO is a population-based stochastic algorithm that exploits a population of individuals representing possible solutions in the search space by simulating the behavior of various swarms like birds, fish, terminates, ants and even human beings. Each particle associated with two vectors: the position vector and the velocity vector to move around the search space. Particles have memory and each particle keeps track of its previous best position. The positions of particles are classified as personal best *pbest* or local best *lbest*, related to a particle only and global best *gbest* which is the best position to a particle with respect to all the particles in the swarm.

   In general, a particle moves in a D-dimensional search space and contains N particles known as swarm size. The position vector and velocity vector of ith particle is represented by $\mathbf{x}_i = (x_{i1}, x_{i2}, \ldots, x_{iD})$ and $\mathbf{v}_i = (v_{i1}, v_{i2}, \ldots, v_{iD})$, respectively. For each particle, its previous best position which is represented by $\mathbf{x}_{pbest} = (x_{pbest1}, x_{pbest2}, \ldots, x_{pbestD})$ while best of all particles in the swarm by $\mathbf{x}_{gbest} = (x_{gbest1}, x_{gbest2}, \ldots, x_{gbestD})$. In literature, the basic equations for updating the velocity and position of the ith particle for the PSO algorithm are presented most popularly in Eqs. (1.14a) and (1.14b), respectively.

$$\mathbf{v}_i(t+1) = \mathbf{v}_i(t) + c_1 r_1(\mathbf{x}_{pbest_i}(t) - \mathbf{x}_i(t)) + c_2 r_2(\mathbf{x}_{gbest_i}(t) - \mathbf{x}_i(t)), \quad (1.14a)$$
$$\mathbf{x}_i(t+1) = \mathbf{x}_i(t) + \mathbf{v}_i(t+1), \quad\quad\quad\quad\quad\quad\quad\quad\quad\quad\quad\quad\quad (1.14b)$$

where t represents the current generation, c_1 and c_2 are two positive constants known as *acceleration coefficients* referred to as cognitive and social parameters, respectively. c_1 and c_2 accelerate the particles towards the personal best and global best positions of the swarm. Adjustment of these coefficients is very important in PSO, which allow particles to move far from the target region for small values whereas high values provide abrupt movement towards or past the target region in the search space [56]. The constants r_1 and r_2 are uniformly distributed random numbers in [0, 1]. The Eqs. (1.14a) and (1.14b), which govern the working principle of PSO for all the particles in the swarm.

A constant, maximum velocity (v_{max}) is used to limit the velocities of the particles and improve resolution of the search space. Large values of v_{max} facilitate global exploration, whereas smaller values encourage local exploitation. If v_{max} is too small, the swarm may not explore sufficiently in the search space while too large values of v_{max} risk the possibility of missing a good region. A most popular variant of PSO was proposed by Shi and Eberhart [57] which incorporates a new parameter ω known as inertia weight factor into the velocity updating Eq. (1.14a). The modified velocity updated equation of ith particle at generation t is presented in Eq. (1.15).

$$\mathbf{v}_i(t+1) = \omega\, \mathbf{v}_i(t) + c_1 r_1 (\mathbf{x}_{pbesti}(t) - \mathbf{x}_i(t)) + c_2 r_2 (\mathbf{x}_{gbesti}(t) - \mathbf{x}_i(t)). \qquad (1.15)$$

The main role of the inertia weight factor ω is to control the magnitude of the velocities and alleviate the swarm-explosion effect that sometimes influence the convergence of the PSO [58] algorithm. From Eq. (1.15), it is clear that the velocity of the ith particle at time instant t, $\mathbf{v}_i(t)$ would be small if $|\mathbf{x}_{pbest_i}(t) - \mathbf{x}_i(t)|$ and $|\mathbf{x}_{gbest_i}(t) - \mathbf{x}_i(t)|$ are small enough and $\mathbf{v}_i(t)$ cannot attain a large value in the upcoming generations. This phenomenon is described as a loss of exploration power and can occur at an early stage of the search process. Moreover, the swarm suffers from loss of diversity in later generations while personal best of particles ($\mathbf{x}_{pbest}$) are close enough to global best of the swarm ($\mathbf{x}_{gbest}$) [58]. The pseudo-code of the standard PSO algorithm is presented in Algorithm 3.

1.7.4.2 Bees Algorithm (BA)

The Bees Algorithm [59] is a swarm intelligence based algorithm inspired by the foraging behavior of honey bees and used for finding a global optimum solution for a given optimization problem. Scout bees i.e. candidate solutions are randomly generated in the search space and the quality of the visited locations have corresponded with the fitness value. The generated solutions are ranked and other bees are recruited in the neighborhood of the highest ranking locations in the search space. This algorithm locates the most promising solutions and selectively explores their neighborhoods looking for the global minimum (for minimization problem) of the fitness function.

Algorithm 3 Particle Swarm Optimization (PSO)

1: Initialize number of particles (N), inertia weight (ω) and acceleration coefficients (c_1 and c_2)
2: Randomly initialize position and velocity V of N particles
3: Initially, local best position ($\mathbf{x}_{pbest}$) of a particle is assigned to initial position of the particle
4: Evaluate fitness values of all the particles
5: Calculate *best particle* in the swarm as $\mathbf{x}_{gbest}$
6: **while** Termination criteria is not satisfied **do**
7:　　**for** $i = 1, 2, ..., N$ **do**
8:　　　　Calculate velocity of ith particle using Eq. (1.14a) or (1.15)
9:　　　　Calculate position of ith particle using Eq. (1.14b)
10:　　　　Evaluate fitness values of each particle
11:　　　　**if** $f(\mathbf{x}_i) \leq f(\mathbf{x}_{pbest,i})$ **then**
12:　　　　　　$\mathbf{x}_{pbest,i} = \mathbf{x}_i$
13:　　　　**end if**
14:　　　　**if** $f(\mathbf{x}_i) \leq f(\mathbf{x}_{gbest})$ **then**
15:　　　　　　$\mathbf{x}_{gbest} = \mathbf{x}_i$
16:　　　　**end if**
17:　　　　Update i^{th} particle position and velocity to $\mathbf{x}_i$ and $\mathbf{v}_i$
18:　　**end for**
19: **end while**

In general, the population contains a N number of scout bees which are randomly scattered with uniform probability distribution across the search space. Therefore, the jth element of the ith solution $\mathbf{x}_i$, x_{ij} is expressed as follows:

$$x_{ij} = x_{ij}^{min} + (x_{ij}^{max} - x_{ij}^{min})\, rand\,(0, 1), j = 1, 2, \ldots, D \qquad (1.16)$$

where x_{ij}^{min} and x_{ij}^{max} denote the lower and upper bound of the jth component of the ith individual of the swarm and D denotes the dimension of any swarm. Each scout bee evaluates the visited site i.e. solution through the fitness function. After initialization of the scout bees, BA enters into a cycle which is composed of four phases [59]. Following phases are executed sequentially until stopping condition is met.

Waggle Dance:

Say, a N number of visited sites are ranked based on fitness information and a B_s number of sites with the highest fitness (minimum measure) are selected for local search. The local search is performed by other bees (foragers) that are directed to the neighborhood of the selected sites. Each scout bees that are returned from one of the B_s best sites performs the 'waggle dance' to recruit nest mates for local search. The scout bees are visited the first elite (top-rated) sites E_s among the B_s sites recruiting E_r bees for neighborhood search. The remaining ($B_s - E_s$) sites that visited by scouts are recruiting $B_r(\leq E_r)$ bees for neighborhood search.

Local Search:

For each of the B_s selected sites, the recruited bees are randomly placed in a neighborhood of the high fitness location marked by the scout bee. This neighborhood is defined as an D-dimensional hyper box of sides $a_1, a_2, \ldots, a_D$, centered on the scout bee. For each neighborhood, the fitness is evaluated by the recruited bees. If one of the recruited bees lands in a position of higher fitness than the scout bee, the recruited bee is treated as the new scout bee. At the end of the local search, only the fittest bee is retained. The fittest solution visited so far is thus taken as a representative of the whole neighborhood.

Global Search:

In the global search phase, $(N - B_s)$ bees are placed according to Eq. (1.16) across the search space for a new solution. Random scouting represents the exploration capability of the BA.

Population Update:

At the end of each cycle, the population is updated from two groups. The first group comprises the B_s scout bees which are associated with the best solution of each neighborhood and represents the results of the local exploitative search. The second group is composed of the $(N - B_s)$ scout bees associated with a randomly generated solution and represents the results of the global exploratory search.

The pseudo-code of the standard bees algorithm is shown in Algorithm 4.

Algorithm 4 Bees Algorithm (BA)

1: Initialize population size and algorithm parameters
2: Randomly generate population and evaluate fitness values of each population member.
3: Sort the initial population based on fitness values
4: **while** stopping criterion not met **do**
5:    *Select* sites for neighborhood search
6:    *Recruit* bees for selected sites (more bees for best elite sites) and evaluate fitness
7:    *Select* the fittest bee from each patch
8:    Assign remaining bees to search randomly and evaluate fitness values
9: **end while**

1.7.5 Physical Phenomena Based Algorithms

1.7.5.1 Harmony Search (HS)

Harmony search (HS) algorithm is a new optimization algorithm that based on the improvisation processes of musicians proposed by Greem et al. [60] in 2001. During the improvisation processes, musicians search for a perfect state of a harmony by slightly adjusting the pitch of their instruments. A better harmony searching process is replicated for finding better solutions to optimization problems. The HS is a population based metaheuristic algorithm where population represents as harmony memory (HM). Besides that there are few control parameters such as harmony memory size (HMS), harmony memory considering rate (MHCR) used for determining whether a decision variable is to be selected from HM or not, pitch adjusting rate (PAR) used for deciding whether the selected decision variable is to be modified or not. The basic HS algorithm consists five phases such as initialization of HM, improvisation of a new harmony, updating of HM and stopping criterion describe as follows:

Initialization of Harmony Memory:

The HM is a solution matrix of size HMS where each harmony represents a D-dimensional solution vector in the search space. In general, the ith harmony is represented by $\mathbf{x}_i = (x_{i1}, x_{x_i2}, \ldots, x_{iD})$ and generated randomly using Eq. (1.17).

$$x_{ij} = x_{ij}^{min} + (x_{ij}^{max} - x_{ij}^{min})\, rand\,(0, 1), \tag{1.17}$$

where $i = 1, 2, \ldots, HMS$, $j = 1, 2, \ldots, D$; x_{ij}^{min} and x_{ij}^{max} are the lower and upper bound of jth parameters of the ith harmony, respectively.

Improvisation of a New Harmony:

After the HM initialization, the HS algorithm enters in a improvisation process of generating a new harmony vector, $\mathbf{y}' = (y_1', y_2', \ldots, y_D')$ from the HM on the basis of memory considerations, pitch adjustments and randomization. In the memory consideration, jth component, y_j' of $\mathbf{y}'$ is selected randomly from $\{x_{1j}, x_{2j}, \ldots, x_{HMSj}\}$ with a probability of $HMCR \in [0, 1]$ which ensures good harmony is considered in the new harmony vector. Components of the new harmony vector are not chosen from the HM according to the $HMCR$ probability, rather are chosen randomly from their possible range with probability $(1 - HMCR)$. The memory consideration and random consideration represent using Eq. (1.18).

$$y_j' = \begin{cases} y_j' \in \{x_{1j}, x_{2j}, \ldots, x_{HMSj}\} & if \quad rand \leq HMCR \\ x_j^{min} + (x_j^{max} - x_j^{min})\, rand\,(0, 1) & otherwise \end{cases}. \tag{1.18}$$

Furthermore, the component in the new harmony vector obtained from the HM is modified using PAR for finding good solutions in the search space. These components are examined and tuned with the probability PAR as follows:

$$y_j' = \begin{cases} y_j' \pm bw\, rand\,(0, 1) & if \quad rand \leq PAR \\ y_j' & otherwise, \end{cases} \tag{1.19}$$

where bw is a bandwidth distance used for variation in the decision variables and $rand(0, 1)$ is a uniformly distributed random number in [0, 1]. In the improvisation processes, each component of the new harmony vector passes through memory considerations, pitch adjustments and randomizations.

Updation of the Harmony Memory:

In order to update the HM, the new harmony vector ($\mathbf{y}'$) is evaluated using the given objective function. If the objective function value $f(\mathbf{y}')$ of $\mathbf{y}'$ is better than $f(\mathbf{x}_{worst})$ of the worst harmony ($\mathbf{x}_{worst}$) stored in the HM, then $\mathbf{x}_{worst}$ is replaced by $\mathbf{y}'$. Otherwise, the generated new vector is ignored.

Stopping Criterion:

The HS algorithm terminates when the stopping condition, maximum number of improvisation is reached. In this work, the improvisation or generation term is referred to a single generation of the search algorithm. The pseudo-code of the standard HS algorithm is presented in Algorithm 5.

Algorithm 5 Harmony Search (HS) algorithm

1: Initialize the HM size (HMS), memory consideration rate (HMCR), pitch adjustment rate (PAR) and Dimension (D)
2: Generate initial HM randomly and evaluate fitness of each harmonics
3: **while** stopping criterion is not satisfied **do**
4:   **for** each harmony $i = 1, 2, ..., HMS$ **do**
5:     **for** $j = 1, 2, ..., D$ **do**
6:       **if** $rand(0, 1) \leq HMCR$ **then**
7:         Choose an existing harmonic randomly
8:         **if** $rand(0, 1) \leq PAR$ **then**
9:           $HM_{i,j}^{New} = HM_{i,j}^{Old} + rand(0, 1)\, bw$
10:         **end if**
11:       **else**
12:         Generate new harmonics randomly
13:       **end if**
14:     **end for**
15:   **end for**
16:   Accept the new harmonics (solutions) if better
17:   Update the HM
18:   Record the best harmony so far
19: **end while**

A summary of the key concepts of widely used metaheuristic algorithms along with the meaning of various metaphors commonly found in the related literature has been provided in Table 1.1.

Table 1.1 Summary of the most representative metaheuristic algorithms

Source of inspiration	Name of algorithm and developers	Metaphor deciphered	Features of the algorithm
Self-referential mutation, derivative-free algorithms like Neelde-Med method	Differential Evolution (DE) by Stron and Price [47]	Population → set of candidate solutions. Vector → decision variables represent a candidate solution. Target vector → predetermined vector in the population. Donor/Mutant vector → a new vector by adding the weight difference between two population vectors to a third vector. Trial vector → a vector generated from donor and target vector using crossover operation. Selection → the trial vector competes against the population vector of the same index	Stochastic direct search and global optimization algorithm designed for non-linear, non-differentiable continuous function optimization
Migration of species among neighboring islands	Biogeography Based Optimization (BBO) by D. Simon [52]	Habitat → a candidate solution. Suitability index variable → decision variables analogous to a gene in GA. Habitat suitability index → measure of the goodness (fitness) of candidate solution. Migration → sharing information among candidate solutions by immigration and emigration rate	Population-based algorithm for continuous and discrete optimization problem. It is not involved reproduction or generation of children as in GA. BBO maintains set of solutions from one generation to the next rather generating new solutions

(continued)

Table 1.1 (continued)

Source of inspiration	Name of algorithm and developers	Metaphor deciphered	Features of the algorithm
Social foraging behavior of animals such as flocking of birds and the schooling of fish	Particle Swarm Optimization (PSO) by Kennedy and Eberhart [54]	Particle $\rightarrow$ a candidate solution. Swarm $\rightarrow$ population of particles associated with a position vector and velocity vector. Velocity vector $\rightarrow$ the directional distance that the particle has covered the search space. *pbest* $\rightarrow$ the best position attained by a particle so far. *gbest* $\rightarrow$ the best location found so far in the entire swarm or some current neighborhood of the current solution	Popular swarm intelligence based algorithm for continuous, discrete and binary optimization problems. In initial phases of the search, wide explorations are observed but reduces in exploration towards the end. Premature convergence occurred in multi-modal landscape due to particles are accelerated towards the swarm's best position
Honey bees foraging and communication	Bees Algorithm (BA) by Pham et al. [59]	Scout bees $\rightarrow$ a candidate solution. Bee colony $\rightarrow$ population of candidate solutions. Flower patch $\rightarrow$ neighborhood of a solution. Nectar $\rightarrow$ fitness of a solution. Waggle dance $\rightarrow$ sharing information about quality of solution obtained by different search. Recruitment procedure $\rightarrow$ to search further the neighborhoods of the most promising solution	The algorithm combines a neighborhood search strategy with global search and can be used for solving both continuous and combinatorial optimization problems. Neighborhood search may waste fitness evaluations considerably
Improvisation process of Jazz musicians	Harmony Search (HS) by Greem et al. [60]	Harmony $\rightarrow$ candidate solution. Musician $\rightarrow$ a decision variable in a candidate solution. Instruments pitch range $\rightarrow$ the bounds and constraints on the decision variables. Aesthetics $\rightarrow$ fitness function	The algorithm designed for solving combinatorial optimization problem, latter extended for continuous optimization problem. Parameters tuning is very crucial while solving multi-modal problems

1.8 Innovation and Research: Main Contribution of This Volume

The remaining chapters of this monograph are as follows:

- **Chapter** 2 presents a literature review of metaheuristic techniques based protein structure prediction, a scope of the work and contributions to the book.
- **Chapter** 3 explains a chaos based random walk algorithm in continuous search space which is used for generating the landscape structure of a problem. Nature of the protein landscape structure in continuous space is also explained.
- **Chapter** 4 provides the structural features of the PSP problem which are determined by analyzing the landscape structure generated by using the quasi-random sampling technique and city block distance. The performance of some of the well-known optimization algorithms is analyzed based on the structural features.
- **Chapter** 5 describes a DE-based metaheuristic algorithm for solving the PSP problem. In the proposed method, parameters of the DE algorithm is controlled adaptively using Lévy distribution. The algorithm is configured in order to overcome the difficulties associated with the PSP problem (discussed in Chap. 4).
- **Chapter** 6 presents four variants of metaheuristic algorithms for protein structure prediction based on two-dimensional and three-dimensional AB off-lattice model. The algorithms are configured in order to enhance the exploration capability and to avoid being trapped into local optima for solving the PSP problem.
- **Chapter** 7 describe a hybridization framework for solving complex multi-modal problem like PSP. The basic PSO and DE algorithm are improved by employing adaptive polynomial mutation and trigonometric mutation. Next, the improved versions are combined to form a hybrid technique for solving protein structure prediction problem.
- **Chapter** 8 concludes the book chapters by summarizing the main points and providing suggestions for future work.

Chapter 2
Metaheuristic Approach to PSP—An Overview of the Existing State-of-the-art

Abstract This chapter provides an overview of the research in protein structure prediction with metaheuristic techniques using AB off-lattice model. The chapter discuses related work covered under the heads of metaheuristic methods, classified into four categories. The scope of work has been outlined and finally, the chapter ends with a discussion of the contributions we have made in this book.

2.1 Introduction

Protein structure prediction from its amino acid sequence is one of the most challenging problems in computational biology. The biological function of a protein is determined from the three-dimensional structure which has great importance to biology, biochemistry, and medicine. The *Anfinsens thermodynamic hypothesis* [18] states that the native structure of a protein corresponds to the global minimum of the free energy surface of the protein, therefore, the protein structure prediction (PSP) problem can be translated into a global optimization problem.

The *ab-initio* prediction approach based on the *Anfinsens thermodynamic hypothesis*, is widely used for PSP. A successful study of *ab-initio* structure prediction entails two key parts. The first part is to state the simplified mathematical model corresponds to the protein energy function, while the second part is to develop a powerful optimization method to search for the global minimum of the potential energy function. The protein energy function is formulated based on the physical model which reveals atomic details of the structure of proteins. The AB off-lattice model provides more atomic interactions than the HP-lattice model. However, the protein energy function based on the off-lattice model is *NP*-hard and complex non-linear optimization problem, solved by the researchers using metaheuristic techniques.

The aim of this chapter is to provide an overview of existing metaheuristic techniques for PSP in both 2D and 3D AB off-lattice model. For the sake of convenience, metaheuristic based *ab-initio* PSP methods are classified into four categories: Evolution based metaheuristic algorithms, Swarm Intelligence based metaheuristic algorithms, Physical phenomena based metaheuristic algorithms and Hybrid algorithms. The chapter then discusses the scope of selecting the metaheuristic algorithms which

© Springer International Publishing AG 2018
N. D. Jana et al., *A Metaheuristic Approach to Protein Structure Prediction*, Emergence, Complexity and Computation 31, https://doi.org/10.1007/978-3-319-74775-0_2

find a solution of PSP problem using fitness landscape analysis. Finally, the chapter explains the major contributions made in the book.

The outline of the chapter is as follows: Sect. 2.2 provides related work on evolutionary algorithms based PSP in both 2D and 3D AB off-lattice model. The literature review on the *ab-initio* PSP based on swarm intelligence, physical phenomena based algorithms are presented in Sects. 2.3 and 2.4. Section 2.5 presents hybrid metaheuristic based algorithm for the PSP problem. The scope of the work and the main contributions are described in Sects. 2.6 and 2.7.

2.2 Evolution Based Metaheuristic Algorithms

The evolutionary algorithm is a population-based metaheuristic optimization algorithm has been developed inspired by biological evolution mechanism. The *ab-initio* approach for predicting the tertiary structure of a protein is nothing but an energy optimization problem, where the energy function may not be differentiable. Therefore, it is impossible to solve the PSP problem using conventional mathematical models. Researchers commonly use evolutionary algorithms to solve the PSP problem.

Parpinelli and Lopes [61] proposed an ecology inspired (ECO) algorithm to solve the PSP problem. The proposed scheme is a type of evolutionary algorithm developed by employing cooperative search strategies where the population of individuals co-evolve and interact among themselves using ecological concepts. The ECO system is modeled in two ways: homogeneous and heterogeneous. A homogeneous model implies that all populations evolve in accordance with the same optimization strategy and configured with the same parameters. Any changes in the strategies or parameters on at least one population characterizes the heterogeneous model. Six different configurations are employed in a homogeneous model. Authors claim that the searching process of a heterogeneous model is more robust than the other approaches due to a proper balance between intensification and diversification strategies. The proposed method experimented only on fewer artificial protein sequences and parameter settings are very limited.

In [62], Olson et al. developed homologous crossover based EA for solving PSP problem. Authors incorporated *ab-initio* features to improve the state-of-the-art in structure prediction method. An initial population is constructed randomly and the population POP_t in each subsequent generation t is obtained by a strategy where all the conformations of the previous population POP_{t-1} are first duplicated, then crossover, mutation, and local search operators are applied for getting the nearby local minimum. To generate child conformations, 1-point, 2-point, and homologous 1-point crossover are employed and added to the population POP_t. The molecular fragment replacement technique is used in mutation operation to improve sampling of near-native conformation over rotating angles, that are uniformly sampled at random. The proposed method is compared with Monte Carlo-based techniques and reveals its effectiveness in *ab-initio* protein structure prediction.

Bošković and Brest [63] proposed differential evolution (DE) variant which is based on self-adaptive DE to improve the efficiency and reduces the number of control parameters for protein folding optimization. The population vector of the proposed method is encoded as bond and torsional angles. The operators are controlled by the parameters like mutation scale factor (F), crossover rate (CR) and population size (NP) which are self-adapted by a mechanism based on jDE [64] algorithm. After performing *DE/best/1/bin* mutation strategy, crossover and selection; temporal locality and reinitialization schemes are incorporated to speed up the convergence rate of the proposed method. Temporal locality is used to calculate a new vector when the trial vector is better than the corresponding vector from the population. The author mentioned that the proposed scheme may be stagnant at local minima and utilizing reinitialization strategy to get rid of the local minima. Authors reported that the proposed algorithm is superior to the other algorithms and obtained lower free energy values.

Kalegari and Lopes [65] proposed two different versions of DE as basic and adaptive DE. A parallel master-slave architecture based on message passing interface (MPI) has been developed in a clustered way to address the PSP problem. Authors implemented a parallel architecture based on MPI to reduce the evaluation cost of the protein energy function. Standard MPI provides a communication and control between the processes running in the different systems with same configurations which divide the computational load and improve overall performance. Master-slave is a topology where master process controls the DE algorithm and distributes individuals to several slave processes to evaluate the energy function. In this architecture, explosion, mirror mutation and adaptation strategies are adopted for better efficiency of the proposed algorithm. An explosion has been occurred by using decimation or mass extinction methods while no improvement can be made on the best solution in a certain number of steps. All bond angles are mirrored in mirror mutation and adaptive DE is incorporated borrowed from the self-adaptive DE algorithm. Authors claimed that the proposed method is adequate and promising for solving the PSP problem. However, several strategic incorporations make the proposed algorithm costlier.

A sequential and parallel differential evolution (DE) algorithm [66] was proposed to predict the protein structure based on 2D AB off-lattice model. The Parallel DE was implemented using master-slave and ring-island concept and experimented on four artificial protein sequences with different lengths within the range 13–55. The results obtained using parallel approach outperform the sequential ones for smaller length (for example 13 and 21) protein instances whereas degrades the performance on a sequence of lengths 34 and 55.

In [67], Sar and Acharyya implemented six variants of GA which are characterized by using different selection and crossover techniques. The proposed variants are employed random, elitist and tournament selection techniques to keep up individuals for Crossover. The proposed variants go through two types of the crossover such as single point and double point crossover. Authors set the control parameter values 0.8 and 0.01 as crossover and mutation probability. Elitist selection with two point crossover variant outperforms compared to other variants suggested by the authors.

Fan et al. [68] proposed an improved biogeography based optimization (BBO) algorithm where the migration model is modified with several habitats. Authors pointed out that the high immigrated habitat affects the immigrating process of nearby habitats, which has been tackled by creating a new migration method. In the proposed method, the original features of $\mathbf{H}_i$ is replaced by mixing the features of $\mathbf{H}_i$ and three different features h_{r_1}, h_{r_2} and h_{r_3} taken from three different habitats according to their emigration rates. The mixture model of jth component of the ith habitat $\mathbf{H}_i$ with coefficients α and β is defined using Eq. (2.1).

$$h_{i,j} = \alpha h_{i,j} + \beta(h_{r_1} + h_{r_2} + h_{r_3}) \tag{2.1}$$

The control parameters such as habitat count $= 30$, max generation $= 1000$, elitism number $= 2$, $\alpha = 1$ and $\beta = 1$ are considered for the experiments. Superior results of the proposed algorithm has been shown while compared with the basic BBO and DE algorithm.

2.3   Swarm Intelligence Based Metaheuristic Algorithms

Swarm intelligence (SI) is the collective behavior of decentralized, self-organized systems which are either natural or artificial. Typically, SI systems consist a population of simple agents interacting locally with one another and with the environment. The inspiration often comes from nature, especially biological systems. The agents follow very simple rules and although there is no centralized control structure dictating how individual agents should behave, local and to a certain degree random interactions between such agents lead to the emergence of "intelligent" global behavior. Recently, SI techniques have been applied to solve the PSP problem.

Cheng-yuan et al. [69] developed a quantum-behaved PSO (QPSO) where the population is divided into elite sub-population, exploitation sub-population, and exploration sub-population to determine the protein structure. In the proposed scheme, the population is divided for balancing local exploitation and global exploration during the search process. Fine turning and exploration strategies are used to faster convergence of the proposed method. Bond angles are used to represent the protein conformation and AB off-lattice model reveals the protein energy function.

Liu et al. [70] proposed a variant of the PSO algorithm which balances between exploration and exploitation during protein folding process. The velocity equation of the proposed algorithm is updated using Eq. (2.2).

$$\mathbf{v}_i(t+1) = k(\mathbf{v}_i(t) + c_1 r_1(\mathbf{x}_{pbest_i}(t) - \mathbf{x}_i(t)) + c_2 r_2(\mathbf{x}_{gbest_i}(t) - \mathbf{x}_i(t))), \tag{2.2}$$

where k is the constriction factor, calculated as:

$$k = \frac{2}{|2 - C - \sqrt{C^2 - 4C}|}.$$

The parameters are set as $c_1 = 2.8$, $c_2 = 1.3$, $C = c_1 + c_2$ and the population size $N = 30$. The proposed algorithm was experimented on fewer protein instances and experimental results reveal the feasibility and efficiency to obtain the ground state energy. Another variant of PSO named EPSO was suggested by Zhu et al. [71] for the PSP problem. In the EPSO method, an inference factor is added to the velocity equation of the particles for flying out from the local optima. The velocity of the particles are updated with an addition of inference factor (I_ϵ) and counter k set to 0 when global best fitness has not been changed for a specific (k) times. The ith inference factor, I_{ε_i} is calculated using Eq. (2.3).

$$I_{\varepsilon_i} = \left(\frac{1}{1 + exp(-a/d_i)} - 0.5 \right) 2v_{max}, \qquad (2.3)$$

where d_i is the Euclidean distance from the current ith particle to global particle, a slope parameter $a = 0.5$ and maximum velocity v_{max}. Authors claimed that experimental results achieved lower energy of proteins better than the other methods.

In [72], Zhang and Li proposed a new architecture that characterized the exploration and exploitation capability of the PSO algorithm. The population of the proposed algorithm is partition into three subpopulations as elitist part, exploitative part, and exploratory part. The elitist subpopulation is obtained by using a Gaussian mutation strategy. The positions of all the particles are updated through the PSO algorithm for exploitative subpopulation. Finally, positions of particles in the exploratory subgroup are generated by using an exploration strategy. Thus, the proposed scheme helps to escape from local minima and provides faster convergence speed. Experiments are conducted only on artificial protein instances of which length varies from 13 to 55. However, this algorithm is very complex due to an involvement of several stochastic terms without proper logic. An improved PSO was proposed in [73] for protein folding prediction. Lèvy distributed probability distribution was employed for updating positions and velocities of the selected worst particles to speed up the convergence rate and jump out from local optima. The proposed algorithm was compared to only basic PSO and experimented on relatively fewer artificial and real protein sequences.

An improved Artificial Bee Colony (ABC) [74] algorithm was proposed for protein structure optimization and experimented on relatively fewer artificial and real protein sequences. An internal feedback strategy enhances the performance of the basic ABC algorithm and brings slightly better results compared to other algorithms. A chaotic ABC [75] algorithm was proposed for PSP. Chaotic search is performed when a food source cannot be improved further within a predetermined number of searches. The proposed methodology improves the convergence rate and preventing stuck at local optima in a case of fewer artificial protein sequences. However, the proposed algorithm is computationally expensive and parameter limit selection can be very difficult. In [76], an ABC variant was proposed for 3D PSP and experimented on fewer and shorter artificial protein sequences. The structure quality was improved by using the convergence information of the ABC algorithm at the execution process. Authors demonstrated that the proposed algorithm outperformed other state-of-the-art approaches on some artificial and real protein instances. In [77], the authors extended their work by employing the same method [76] on thirteen real protein sequences and compared the results with the existing results available in the literature. The chaotic ABC algorithm also involves considerable computational overhead.

A comparative study of swarm intelligence algorithms was performed on protein folding problem based on 3D AB off-lattice model [78]. Authors analyzed the performance of the standard versions of the algorithms only on fewer and short artificial protein sequences. The comparison could be more realistic on consideration of real protein sequences than artificial protein sequences.

2.4 Physical Phenomena Based Metaheuristic Algorithms

The metaheuristics which mimic various physical phenomena like annealing of metals, musical aesthetics, memory based have been applied to solve the PSP problem. Lin and Zhu [79] introduced an improved tabu search (TS) for protein structure optimization. The proposed method has been improved by an adaptive neighborhood strategy developed based on annealing process. The authors provide the initial temperature (10^2), the terminal temperature (10^{-7}), the drop rate of the temperature (0.96), the neighborhood size (100) and the candidate solution size (100) for the experiment. Fewer artificial or Fibonacci sequences that do not reveal the efficiency of the proposed algorithm. The authors extended the same method by employing some real protein instances [80]. The heuristic initialization mechanism, the heuristic conformation updating mechanism and the gradient method were employed on the basic Tabu Search (TS) to improve the TS performance on artificial and two real protein sequences [81]. Further, authors utilized a local search mechanism for preventing stuck at local optima and results are quite promising in terms of lowest energy based on 2D AB off-lattice model. However, the proposed algorithm is computationally slow.

The standard harmony search (HS) algorithm, named population-based harmony search (PBHS) was introduced by Scalabrin et al. [82]. The proposed strategy is developed by using graphical processing units (GPU) which provide multiple function evaluations at the same time. GPU performed a task in a way of parallel execution. The population of PBHS includes two type of population: temporary harmonies, named musical arrangement (MA) and harmony memory (HM). The two set HM and MA are taken in the improvisation step and keep the best harmony to form HM at the end of each cycle.

A locally adjusted TS algorithm was introduced in [83] for PSP on both the 2D and 3D AB off-lattice model. Simulated annealing was used to modify the TS neighborhood structure for obtaining faster convergence. The authors claimed an improvement of results compares to the available literature. Choosing of the initial and end temperatures is a great challenge.

2.5 Hybrid Algorithms

Sometimes one optimization technique is not sufficient enough to solve a problem. Then, two or more optimization techniques are merged to get the benefits from all the techniques. The present section describes such hybrid strategy to resolve the PSP problem.

Mansour [84] developed a hybrid algorithm which combined GA and tabu search (TS) algorithm for solving PSP problem. The author utilized TS instead of mutation operator in GA for preventing premature convergence and providing faster convergence rate. The proposed algorithm adopted by a variable population strategy to keep the diversity of the problem. Arithmetic crossover is used in GA and after the population is modified by two TS schemes such as tabu search recombination (TSR) and tabu search mutation (TSM). TSM strategy accepts the worst solution as the current solution to make the algorithm have better hill-climbing capability and strong local searching capability. TSR strategy is used to limit the frequency that the off springs with the same fitness appear to keep the diversity. However, this algorithm can be computationally costlier, especially on longer protein instances due to the involvement of TS strategy.

Zhang et al. [85] proposed a hybrid algorithm where crossover and mutation operator of GA are improved by using TS. The core of the proposed algorithm employed by some strategies such as variable population size, ranking selection, tabu search mutation (TSM) and tabu search recombination (TSR) for maintaining diversity in the solutions and avoid premature convergence. The hybrid algorithm tested on fewer artificial and real protein instances and experimental results indicate superiority of the proposed method, however, computational cost may be increased due to several strategy incorporations.

In [86], Zhon et al. introduced a hybrid optimization algorithm which combines PSO, GA, and TS. In the proposed method each algorithm is improved itself by using some improvement strategy. The velocities of particles are updated by adding an extra stochastic disturbance factor as $\delta_f \, rand(0, 1)$ (δ_f is a very small constant). GA is improved by linear combination applied to crossover operator whereas stochastic mutation strategy is incorporated in TS to make search efficiently. The proposed scheme does not mention how to maintain the velocity of the particles during the execution of improved GA and TS search. Another hybrid technique, named GAPSO suggested by Lin and Zhang [87] which combines GA and PSO to predict the native conformation of proteins. In particular, authors utilized the crossover and mutation strategy concept in PSO. A local adjustment strategy is applied on the global Best practice when it is not updated predefined times.

2.6 Protein Structure Prediction Using Metaheuristics: A Road Map of This Volume

Proteins are the building blocks of life. They perform a variety of tasks such as catalyzing certain processes and chemical reactions, transporting molecules to and from the cell for the preservation of life. The function of an individual protein depends on its three-dimensional structure or more simply its shape. The functional properties play important role in biological science, medicine, drug design, disease prediction, therefore, the structure prediction of a protein is an essential task in the present scenario. Unfortunately, most of the protein structures are predicted using experimental techniques which are very expensive, costly and lengthy process. To overcome such limitations, researchers and scientists are developed computational techniques for predicting the protein structure.

In *ab-initio* or *denovo* protein structure prediction technique, the structure is predicted with a sequence, not similar to any of the proteins with a determined structure in the database i.e. free from template-based modeling. The *ab-initio* method based on *Anfinsen's thermodynamic hypothesis* states that a native structure consumes a global minimum energy. Therefore, for the lowest energy state, the entire conformational space is searched. The protein energy function is formulated based on a physical model in which model provides much higher atomic interactions within a protein sequence. The energy function consumes minimum energy reveals a stable or native structure of a protein. The AB off-lattice model expresses atomic interactions which is very useful for drug design and reason for considering the off-lattice model for PSP problem. However, the protein energy function based solution using off-lattice model is *NP*-hard problem and complex non-linear optimization problem. Therefore, to solve the PSP problem using metaheuristic algorithm is a popular choice to solve the problem.

The aforementioned research studies (discussed in the Sects. 2.2–2.5) mainly focused on developing algorithms and/or their variants for solving the protein structure prediction problem considering relatively fewer and smaller lengths of protein sequences. The reported approaches did not discuss the appropriateness of the methods while solving the PSP problem. Therefore, the approaches lacking a clear understanding of why one method performs well or poorly compare to the others. In general, the problem characteristics or structural features are related to the complexity of a problem and influence the performance of an algorithm. Hence, a fundamental question arises: how to select and configure a best-suited heuristic for solving a specific problem? Fitness landscape analysis (FLA) [88, 89] has been applied successfully for a better understanding of the relationship between the problem characteristic and the algorithmic properties.

Fitness landscape analysis (FLA) is a technique to determine the characteristic features of a problem based on which the most appropriate algorithm is recommended for solving the problem. Since the AB off-lattice model of protein structure prediction is defined in the continuous domain. Thus, the discrete landscape analysis is infeasible for the PSP problem. Therefore, continuous fitness landscape analysis requires an appropriate landscape structure for characterizing the continuous search space. Random walk (RW) algorithm is used for generating the sample points in the search space and landscape structure is created based on the relative fitness of the neighboring sample points. Existing information and statistical techniques for analyzing the fitness landscape structure are adapted to determine the structural features of the PSP problem.

The proper choice of control parameters and operator configurations are very important for the performance of an optimization algorithm while solving a problem. We developed a mechanism to select the control parameters for widely used differential evolution (DE) algorithm with a view to solving PSP like multi-modal problems. PSP problem based on 2D and 3D AB off-lattice model is an NP-hard complex optimization problem. Moreover, the problem consists difficult features such as ruggedness, multi-modality, deceptiveness and multiple funnels reflecting the performance of any metaheuristic optimization algorithm. Proper configurations are needed in the operators of a metaheuristic algorithm design to develop an efficient algorithm for multi-modal optimization problem like PSP problem.

Hybridization is a technique which combines two or more algorithms in an efficient way to solve such type of problems.

2.7 Potential Applications Areas for the Metaheuristics Based Structure Prediction

The major contributions made in this book are as follows:

- Random Walk (RW) algorithm has been used for generating sample points in the search space and fitness landscape structure is created based on the relative fitness of the neighboring sample points. A new algorithm for random walk in multi-dimensional continuous space has been developed, called Chaotic Random Walk (CRW) algorithm [90], described in Chap. 3. This approach is used to sample spaces when information on the neighborhood structure is available and used as the basis of generating landscape structure. It has been observed that the proposed walk provides better coverage of a search space than a simple random walk algorithm.
- Another random walk algorithm has been developed based on the quasi-random sampling technique and city block distance to generate protein landscape structure [91]. The performance of some of the well-known optimization algorithms is analyzed based on the structural features as an illustration of the usefulness of the former research agenda. This work is described in Chap. 4. Comprehensive simulations are carried out on both artificial and real protein sequences in 2D and 3D AB off-lattice model. Finally, suggested the most appropriate algorithms for solving the PSP problem.
- A new variant of the DE algorithm has been proposed, called Lèvy distributed DE (LdDE) [92] to avoid the difficulties associated with the PSP problem discussed in

Chap. 4. Experiments are conducted on complex benchmark test functions as well as real protein sequence to validate the proposed method and described in Chap. 5.

• The PSP problem consists difficult features such as ruggedness, multi-modality, deceptiveness and multiple funnels for both 2D and 3D AB off-lattice model (as discussed in Chap. 4) that reflects on the performance of any metaheuristic optimization algorithm. PSO with local search, called PSOLS [93] has been presented to prevent stuck at local optima for predicting protein structure using 2D AB off-lattice model and described in Section of Chap. 6.

• An adaptive polynomial mutation (APM) based BA (APM-BA) [94] has been developed to enhance diversity in the solution space for solving the PSP problem based on 2D AB off-lattice model. The APM strategy finds solutions to jump out from the local optima in the search landscape. This approach described in Section of Chap. 6.

• Mutation operator of the BBO algorithm is configured by chaotic systems and developed a chaotic mutation based BBO, called BBO-CM [95] and presented in Section of Chap. 6. The performance of the developed method is tested on artificial and real protein instances based on 3D AB off-lattice model.

• The basic HS is a metaheuristic algorithm not too efficient for solving a multi-modal problem like PSP problem. The HS algorithm is configured by using difference mean based perturbation, called DMP-HS [96] has been developed and described in a Section of Chap. 6. Experiments are carried on real protein instances and shown superior performance by faster convergence.

• An integrated hybrid framework for solving the complex PSP problem has been developed, called hybrid PSO and DE (HPSODE) [97] which combines improved PSO and improved DE and described in Chap. 7. Experiments are performed and validate the performance of the proposed algorithm on ten real protein sequences with different lengths. This approach provides better results compared to the other hybrid techniques of PSO and DE.

Chapter 3
Continuous Landscape Analysis Using Random Walk Algorithm

Abstract This chapter describes a chaos based random walk (CRW) algorithm for analyzing landscape structure in continuous search spaces. Unlike the existing random walks, no fixed step size is required in the proposed algorithm, rather conduct the random walk. The chaotic map is used to generate the chaotic pseudo random numbers (CPRN) for determining the variable-scaled step size and direction. The superiority of the new method has been demonstrated while comparing it with the simple and progressive random walk algorithms using histogram analysis. The performance of the proposed CRW algorithm is evaluated on the IEEE Congers on Evolutionary Computation (CEC) 2013 benchmark functions in continuous search space having different levels of complexity. The proposed method is applied to analyzing the landscape structure for protein structure prediction problem in continuous search space.

3.1 Introduction

Optimization algorithms developed using stochastic and heuristic rules have been applied successfully to solve a wide variety of real-world problems. Despite many success stories of these algorithms, it is not known a priori that which algorithm or algorithmic variants are in general more suitable for a particular kind of problems [98, 99]. It is equally true that not a single heuristic algorithm can perform equally well on all possible optimization problems [100]. Current state-of-the-art algorithms based on evolutionary mechanism or nature inspired metaheuristics do not provide a clear mapping between the problem complexity and the available algorithms. The complexity of a problem and the ability of a heuristic search technique to obtain best optimal solution can be determined by using the fitness landscape analysis (FLA) [101, 102].

A landscape can be interpreted as a surface in the search space that defines neighborhood relation among the potential solutions. FLA detects and highlights salient features of an optimization problem such as ruggedness, modality, deception,

© Springer International Publishing AG 2018
N. D. Jana et al., *A Metaheuristic Approach to Protein Structure Prediction*, Emergence, Complexity and Computation 31, https://doi.org/10.1007/978-3-319-74775-0_3

presence of funnels, smoothness, neutrality and variable separability [103]. A comprehensive survey of different features for an optimization problem can be found in [88]. In the literature, FLA is usually applied to the problems represented as bit strings in discrete search space, where Hamming distance is used to generate successive points of a random walk (RW) on the landscape. The techniques such as fitness distance correlation [104], the density of states [105], dispersion metric [106] and exploratory landscape analysis [107] have been employed for a simple random sampling of solution points on the discrete search space to characterize the problem by analyzing the fitness landscape structure.

RW algorithms are developed to capture a sequence of neighboring interaction of sample points over the search space. FLA is performed on the sequence of neighboring sample points using measures like autocorrelation, correlation length [108] and an entropy measure of ruggedness, smoothness with respect to neutrality [109, 110]. Two sampling strategies are usually applied as the online and off-line basis. In online sampling, sample points are generated by a search algorithm itself [111] over the search space. A particular search operator is employed to define the neighborhoods in the search space for off-line sampling [112, 113]. RW is a kind of off-line sampling to generate a sequence of neighboring sample points on the search space without using any fitness information (fitness function or objective function) of an optimization problem. It is very important to notice that features such as ruggedness, modality, deception, funnels, smoothness or neutrality are not the features of the fitness function, rather the features of a fitness landscape structure over a search space. The same fitness function of an optimization problem can generate many different landscapes depending on the sampling points explored by a searching algorithm or a search operator [114]. So, different fitness landscapes are produced by exploring multi-dimensional search space in different ways especially in context to a continuous optimization problem. Hence, different observations are made on the features such as ruggedness, deception or neutrality. Therefore, the importance of the RW algorithms or sampling techniques is associated with FLA which provides information about the landscape structure on the search space.

We proposed the Chaotic Random Walk (CRW) algorithm [90] in continuous search space to obtain the landscape structure for determining the problem characteristics. CRW is an outcome based on the inspiration of the chaotic behavior in connection with generating a pseudo-random number (PRN) deterministically. The idea is to utilize CPRNs, given by chaotic map generating sample points of the RW. The CPRNs control the length of the step size and determine the direction of the proposed algorithm. The walk starts at a random position of the search space and the walk length increases with perturbation of the neighboring points. The current position of the walk is perturbed by the product of CPRN (δ) and the step size. Importantly, the perturbation component determines the variable step size and direction in each dimension for the current position of the walk. Based on the dynamics of the CPRNs generated by the Chebyshev and iterative with infinite collapse chaotic maps, two CRW algorithms are proposed, namely, Chebyshev chaotic random walk

(CCRW) and iterative chaotic random walk (ICRW). Through the histogram analysis, we show that the proposed CCRW and ICRW provide the sample points which are uniformly distributed over the search space. The ruggedness and deception are quantified efficiently for all the IEEE CEC 2013 benchmark functions landscape structure generated by the CCRW and ICRW algorithms. Additionally, we measure ruggedness often selected functions in different dimensions (D) on the basis of the proposed random walk algorithms and results are compared with a simple random walk (SRW) and progressive random walk (PRW) [115]. The objective of the work is to describe the chaos based random walk algorithm and illustrate how it works in FLA for measuring the ruggedness of an optimization problem in continuous search space.

The chapter is organized as follows: Sects. 3.2, 3.3, 3.4 and 3.5 and describes basic fundamentals of random walk, chaotic maps, and fitness landscape analysis measures. Section 3.6 explains the proposed methodology for analyzing the continuous search space based on fitness landscape analysis techniques. Experimental results and the corresponding discussions are reported in Sect. 3.7 considering a number of continuous benchmark problems with different characteristics and different level of dimensions. Landscape structure of real protein sequences is also investigated by the proposed method, presented in Sect. 3.8. Finally, a summary of this chapter is presented in Sect. 3.9.

3.2 Simple Random Walk (SRW)

A random walk is the mathematical formalization of a path consisting of successive random steps. A random walk has been applied in diverse disciplines for explaining the behaviors of the processes and served as a fundamental model for the stochastic activity. Various types of random walk models are defined for combinatorial and continuous optimization problems. The random walk was first introduced by Karl Pearson in 1905 [116] and described as: *"A man starts from a point O and walks l yards in a straight line; he then turns through any angle whatever and walks another l yards in a second straight line. He repeats this process n times"*. The random walk may be isotropic or anisotropic. In an isotropic walk, movement of the locations is identical in all directions, therefore, invariant with respect to the direction. Due to an imposition of restrictions or limitations on the step size or on the directions, the walk would be an anisotropic one. The directed random walk in continuous space [117] is described as a model of random motion with preferential direction. Probability distribution such as uniform, Gaussian or Lévy is used to determine the length of a step for a random walk algorithm. Recent investigation [118] reveals that the choice of step size plays a crucial role in a random walk algorithm.

The simple random walk is constructed by sampling the starting point $\mathbf{x}_1$ randomly and then by generating the next points $\mathbf{x}_2, \ldots, \mathbf{x}_N$ until a predefined number of points (N) in the neighborhood is achieved. For the multidimensional continuous search space, position of each dimension increases or decreases by a step of random size

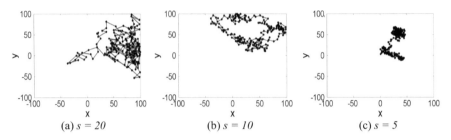

(a) $s = 20$ (b) $s = 10$ (c) $s = 5$

Fig. 3.1 Three independent simple random walks of 50 steps each with variable step size of simple random walk algorithm

and direction. The neighborhood definition always ensuring that the walk stays within the bounds of the search space. The general framework of an SRW is described in Algorithm 6. Figure 3.1 shows the position of sample points generated by an SRW in a two-dimensional space using different values of the step size (s). It has been observed that the steps in random directions have a tendency to cluster the points in limited areas of the search space which is precisely clear when the step size (s) decreases. This phenomenon causes a poor coverage of the search space.

Algorithm 6 SRW Algorithm

1: Initialize dimension D of the problem and domain for each dimensions $[x_1^{min}, x_1^{max}]$, $[x_2^{min}, x_2^{max}]$,...,$[x_D^{min}, x_D^{max}]$. Specify the number of steps (N) and the bound on the step size (s) in the walk
2: Create a matrix M for storing the walk of size N with D-dimension
3: Generate a random position within the domain of the problem
4: M[$1, D$]=$(x_{11}, x_{12}, ..., x_{1D})$
5: **for** $i = 2$ to N **do**
6:    **for** $j = 1$ to D **do**
7:       Generate a random number, r in the range $[-s, s]$
8:       M[i, j] = M[$i - 1, j$] + r
9:       **if** M[i, j] > x_j^{max} **then**
10:         M[i, j]= x_j^{max}
11:       **end if**
12:       **if** M[i, j] < x_j^{min} **then**
13:         M[i, j] = x_j^{min}
14:       **end if**
15:    **end for**
16: **end for**

3.3   Random Walks for Fitness Landscapes Analysis

A fitness landscape is constructed by assigning a fitness value to each point in the search space. For a binary problem, each point is represented by a string of bits having a finite number of neighboring points. Neighboring points are obtained by using Hamming distance. A random walk could be implemented [110] by starting from a randomly chosen point, generating all neighbors of the current point by mutation (bit flip), choosing randomly one neighbor as the next point, and then generating all neighbors of the new point and so on. Fitness landscape analysis techniques on binary landscapes were proposed using autocorrelation measures [108], entropic measures [109, 110, 119] and based on the length of neutral walks [120]. A recent contribution [121] discussed the landscape of combinatorial optimization problems using a random walk over the binary landscape. The work [121] emphasized comparison of the fitness landscape using four classical optimization problems: max-sat, graph coloring, a traveling salesman and quadratic assignment problem with respect to local optima.

However, in the case of continuous search spaces, there is no equivalent set of all possible neighbors of any point as in the binary landscape, since the number of neighbors of a point is theoretically infinite. Morgan and Gallagher [122] proposed length scale property for continuous problems based on random walks. This approach characterizes continuous landscapes based on the entropy of length scale values sampled by using a Lévy flight. In [123], the information content method was extended to continuous fitness landscapes. An adaptive random walk with variable step size was proposed where biased sample and additional uncertainty introduced by the distance between observations were not considered. In [115], a progressive random walk algorithm was proposed for continuous fitness landscape analysis, estimating macro ruggedness of a function based on this random walk. Iba and Shimonishi [124] proposed a chaotic walk to generate a pattern on a two-dimensional plane using a chaotic map. The walk turns around at the angle calculated by the logistic map dynamics and applied to system diversification. Our proposed approach is different from this method and applied to fitness landscape analysis.

All of the fitness landscape techniques discussed above are based on random walks, created by generating sample points in a sequence based on some concept of the neighborhood.

3.4   Chaotic Map

Often the Pseudo Random Numbers (PRNs) are generated based on some parametric probability distribution. For generating unbiased randomness, researchers have been relying on chaotic patterns which can emerge from even a simple deterministic dynamics. Chaos is a non-linear deterministic approach which generates PRNs, and its dynamical behaviors are highly sensitive to initial conditions, and ergodicity. In chaos theory, a chaotic map is used to generate a sequence of pseudo-random numbers known as chaotic pseudo random numbers (CPRN). One dimensional chaotic maps are widely used to generate chaotic behaviors in PRN [125]. Some well known one-dimensional maps which yield chaotic behaviors are outlined below [126]:

3.4.1 Logistic Map

Logistic map [127] is a polynomial mapping and can produce random numbers with chaotic characteristics. It provides dynamic characteristics without any random disturbance and is defined by

$$x_{i+1} = \lambda x_i (1 - x_i), \tag{3.1}$$

where x_i is the ith chaotic number lying in $(0, 1)$. λ is a parameter and most frequently set as $\lambda = 4$.

3.4.2 Tent Map

Tent map and logistic map [128] are conjugate to each other. It has specific chaotic effects and generates chaotic sequences in the range $(0, 1)$. The tent map is defined as:

$$x_{i+1} = \begin{cases} x_i/0.7, & x_i < 0.7 \\ (10/3)x_i(1 - x_i), & x_i \geq 0.7 \end{cases}. \tag{3.2}$$

3.4.3 Chebyshev Map

Chebyshev map [125] generates chaotic sequence in symmetrical regions and lies in the range $[-1, 1]$. The map is defined as:

$$x_{i+1} = cos(5cos^{-1}(x_i)). \tag{3.3}$$

3.4.4 Cubic Map

Cubic map is one of the most commonly used maps generating chaotic sequences in various applications like cryptography. The sequences are obtained in the following way:

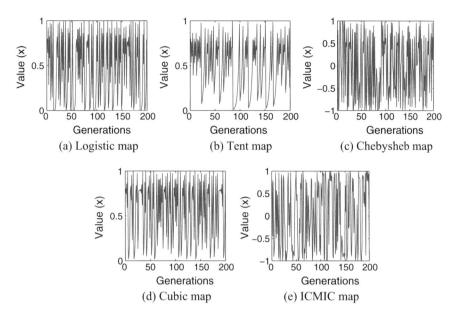

Fig. 3.2 Different chaotic behaviors of chaotic maps

$$x_{i+1} = 2.59x_i(1 - x_i^2), \tag{3.4}$$

where x_i is the ith CPRN. It generates the sequence of random numbers in the range 0 to 1.

3.4.5   ICMIC Map

Iterative chaotic map with infinite collapse (ICMIC) [127] generate chaotic sequences in $(-1, 1)$ and defined in Eq. (3.5).

$$x_{i+1} = sin(70/x_i). \tag{3.5}$$

The chaotic behavior of these chaotic maps in one dimension space with 200 generations are shown in Fig. 3.2. Here, the initial chaos variable is set to 0.8 for each chaotic map. In Fig. 3.2, it can be observed that the chaotic dynamics or ergodic property of different chaotic maps are different.

3.5   Landscape Measure Indices

3.5.1   Entropy Based Measure

The structures of fitness landscape influence the search moves of an Evolutionary Algorithms (EAs) try to locate the global optimum. Therefore, problem hardness could be measured by the structure of the fitness landscape. There are various characteristics that define the structure of fitness landscape, such as modality, epistasis, ruggedness and deceptiveness [101]. Rugged search space refers to the presence of several peaks and valleys. Heavily rugged fitness landscapes are difficult to search compared to a smooth landscape [102]. Moreover, a variety of shapes on the landscape associated with the local neighborhood of a landscape is measured by information content. In [109, 110], Vassilev et al. proposed an information theoretic technique where the information content (entropy measure) is used as a measure of ruggedness of landscape. For estimating ruggedness, a sequence of fitness values $\{ f_1, f_2, ..., f_N \}$ are obtained by a random walk of N steps on a landscape over the search space. The sequence is a time series representing a path on the landscape and contains information about the structure of the landscape [129].

A time series is represented as a string, $S(\varepsilon) = s_1 s_2 s_3 ... s_n$, where symbols $s_i \in \{\bar{1}, 0, 1\}$ are obtained by the function shown below:

$$s_i = \Psi_{f_k}(i, \varepsilon) = \begin{cases} \bar{1} & if \ f_k - f_{k-1} < -\varepsilon \\ 0 & if \ |f_k - f_{k-1}| \leq \varepsilon \ , \\ 1 & if \ f_k - f_{k-1} > \varepsilon \end{cases} \tag{3.6}$$

where parameter ε is a real number that determines structure of the landscape obtained using the string $S(\varepsilon)$. Small ε value provides a high sensitivity to the differences between neighboring fitness values. Based on the above formulation of the string $S(\varepsilon)$, an entropic measure $H(\varepsilon)$ is calculated using Eq. (3.7).

$$H(\varepsilon) = -\sum_{p \neq q} P_{[pq]} \log_6 P_{[pq]}, \tag{3.7}$$

where p and q are elements of the set $\{\bar{1}, 0, 1\}$, and $P_{[pq]}$ is calculated as:

$$P_{[pq]} = \frac{n_{[pq]}}{n}, \tag{3.8}$$

where $n_{[pq]}$ is the number of occurrences of the sub-blocks pq in the string $S(\varepsilon)$. The value of $H(\varepsilon)$ lies in the range [0, 1] which is an estimate of the variety of shapes in the landscape. In Eq. (3.7), all possible sub-blocks pq $(p \neq q)$ of length two in the string $S(\varepsilon)$ represent elements that are rugged. The base of the logarithmic function is 6 making a provision for 6 possible rugged shapes in the landscape. For each rugged element, Eq. (3.8) calculates the probability of occurrence of that element. Therefore, higher the value of $H(\varepsilon)$, more the variety of rugged shapes in the walk, resulting more rugged landscape structure.

Sensitivity of differences in the fitness values is determined by the parameter ε. The smallest value of ε, say ε^* is called the information stability for which the landscape becomes flat by evaluating $S(\varepsilon)$, a string of 0s. The value of ε^* is calculate using Eq. (3.9) and it determines the scale of the problem and the largest change in fitness likely to be encountered.

$$\varepsilon^* = min\{\varepsilon, s_i(\varepsilon) = 0\}. \tag{3.9}$$

3.5.2 Fitness Distance Correlation (FDC)

Jones et al. [104] proposed FDC to study the complexity of a problem. It determines a correlation between the fitness value and respective distance to the global optimum in the search space. In the landscape, distances between sample points are evaluated using Euclidean distance. In the fitness-distance analysis, the position of the global minimum is assumed to be known a priori. The fitness distance correlation coefficient r_{FD} is calculated using Eq. (3.10).

$$r_{FD} = \frac{1}{s_f s_d N} \sum_{i=1}^{N} (f_i - \bar{f})(d_i - \bar{d}), \tag{3.10}$$

where d_i is the minimum distance of the ith point $\mathbf{x}_i$ to the global minimum $\mathbf{x}^*$. $\bar{d}$ and s_d are the mean and standard deviation of the distances of the sample points to the global minimum. The mean and standard deviation of the fitness values over N sample points are $\bar{f}$ and s_f respectively. The coefficient r_{FD} near to 1 represents the global convex and single funnel landscape whereas 0 value represents a needle-like a funnel in the landscape structure. A negative value of r_{FD} indicates 'deceiving' landscape which misleads the search processes away from the global optimum. Thus, FDC reflects the funnel structure and deception in the search landscape for a problem.

3.6 The Chaotic Random Walk (CRW) Algorithm

Fitness landscape analysis is performed on landscape structure of a problem to evaluate inherent characteristics of the problem. A landscape structure is generated by using a random walk algorithm. RW algorithm is created on the basis of a sequence of neighboring points in the search space. So, FLA highly depends on RWA as well as in which way the sequence of the neighboring points are generated. The goal is to develop an RWA which properly covers the whole search space in such a way that the functional properties of the given problem in any given region can be detected and well characterized by the RWA. The coverage area obtained by the simple random walk algorithm (SRWA) in the search space are limited and inadequate for landscape analysis. We propose a chaotic random walk algorithm (CRW) where CPRN provides the direction and scale of the step size of the walk. The output of the chaotic map is used for generation of the real numbers, applied for creating the random walk.

3.6.1 Chaotic Pseudo Random Number (CPRN) Generation

The basic concept of CPRN is to replace the default PRN with the discrete chaotic map. A chaotic map starts with a random number which is initialized using the default PRN. The next CPRN of the chaotic map is generated by using the previous number as the current number. There are two possible ways to generate CPRN by using a chaotic map [125]. Firstly, CPRN is generated and stored as a long data sequence in the memory during the generation processes. In the second approach, a chaotic map is not reinitialized and any long data series is not stored during the experiment. It is advantageous to keep the current state of the map in memory to obtain the new output values. In our work, the second concept is used to generate CPRN in a walk. We select three chaotic maps for generating CPRN in the experiments such as tent maps, Chebyshev map, and ICMIC map.

3.6.2 Chaos Based Random Walk Algorithm

A D-dimensional point in the continuous search space can be expressed as $\mathbf{x} = \{x_1, x_2, ..., x_D\}$ and range of the search space for each dimension is $[x_1^{min}, x_1^{max}]$, $[x_2^{min}, x_2^{max}],...,[x_D^{min}, x_D^{max}]$, where x_j^{min} and x_j^{max} represent the lower and upper bounds of the variable x_j ($j = 1, 2, ..., D$). The starting point, $\mathbf{x}_1 = \{x_{11}, x_{12}, ..., x_{1D}\}$ of the walk is obtained randomly within the search space as:

$$x_{1j} = x_{1j}^{min} + (x_{1j}^{max} - x_{1j}^{min})rand(0, 1), \qquad (3.11)$$

where $j = 1, 2, ..., D$ and $rand(0, 1)$ is a uniformly distributed random number in [0, 1]. After initializing the starting point as a current point, next point which is the next step of the walk obtained by perturbing each dimension of the current point. In each generation, CPRN is generated for each dimension using a chaotic map and multiplied with step size. Sign of the CPRN determines whether the resulting product is added or subtracted to each dimension of the current point. Moreover, the resulting product provides the step size (s) of the walk. Finally, the product is used to perturb the current point in order to obtain the next point. Next point in the walk is generated recursively as:

$$x_{i,j} = x_{i-1,j} + \delta s, \qquad (3.12)$$

where $i = 1, 2, ..., N$, N denotes number of points in the walk, $j = 1, 2, ..., D$ and $\delta \in (0, 1)$ or $(-1, 1)$ based on the chaotic map. If a point exceeds the search space boundary then it is reflected into the feasible range with exceed amount. The CRW, generated based on the CPRN using tent map is known as the tent chaotic random walk (TCRW). Similarly, Chebyshev and ICMIC chaotic maps generate Chebyshev chaotic random walk (CCRW) and ICMIC chaotic random walk (ICRW). The general framework of the proposed CRW algorithm is presented in Algorithm 7.

Algorithm 7 CRW Algorithm

1: Initialize dimension D of the problem and domain for each dimensions $[x_1^{min}, x_1^{max}]$, $[x_2^{min}, x_2^{max}]$,....,$[x_D^{min}, x_D^{max}]$. Specify the number of steps (N) and the bound on the step size (s) in the walk
2: Create a matrix M for storing the walk of size N with D-dimension
3: Generate starting point of the walk using Eq. (3.11)
4: Create one-dimensional array **A** of size D with default PRN for the initial value of a chaotic map
5: Create one-dimensional array **C** of size D for storing CPRN in each step of the walk
6: **for** $i = 2$ to N **do**
7:   **for** $j = 1$ to D **do**
8:     Generate chaotic PRN $C(j)$ (δ) using Eq. (3.2) or (3.3) or (3.5)
9:     $x_{ij} = x_{i-1j} + \delta * s$
10:    **if** $x_{ij} \geq x_j^{max}$ **then**
11:      $x_{i,j} = x_{ij} - |x_j^{max} - x_{ij}|$
12:    **end if**
13:    **if** $x_{i,j} \leq x_j^{min}$ **then**
14:      $x_{i,j} = x_{ij} + |x_j^{min} - x_{ij}|$
15:    **end if**
16:   **end for**
17:   **A** = **C**
18: **end for**

Generally, an RW algorithm is highly sensitive to the selection of step size. Large step size provide a stochastic jumping to random places in the search space. Therefore, large step size makes no sense to use such a walk to estimate landscape measures such as ruggedness, FDC because adjacent points on the walk are frequently in very different regions of the landscape structure. On the other hand, small step size causes the walk to remain confined in a very small region of the search space which gives lack of information about landscape structure of the whole search space for a portion. Thus, large or small step size in an RW algorithm is inadequate for fitness landscape analysis. Therefore, the step size is to be selected in such a way so that a walk should maintain a reasonable close proximity between the points on the walk and thus should provide a better coverage of the search space. Three CRW algorithms of 200 steps with different step sizes such as 20, 10 and 5 (10, 20 and 40% of the search range along a single dimension) are depicted in Figs. 3.3, 3.4 and 3.5 respectively.

It can be seen that a step size of 20 provides a better coverage area in the search space than smaller step size. Moreover, we noticed in Fig. 3.3 that tent chaotic random

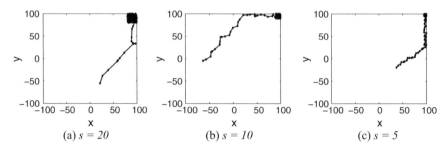

Fig. 3.3 One independent sample walk of 200 steps with variable step size using tent chaotic random walk

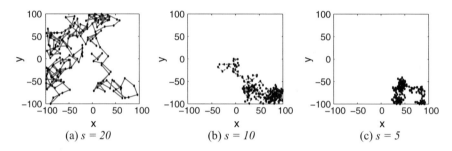

Fig. 3.4 One independent sample walk of 200 steps with variable step size using CCRW

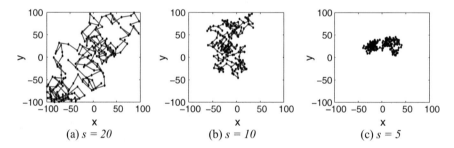

Fig. 3.5 One independent sample walk of 200 steps with variable step size using ICRW

walk (TCRW) is clustered in a very small region of the search space for different step sizes. Tent chaotic map generates CPRN which is in the range (0, 1) and always producing positive perturbation in each dimension of the point and in each step of the walk. Therefore, the points generated using the TCRW are clustered in the upper right corner of the search space. Figures 3.4 and 3.5 provide better coverage area obtained by the CCRW and ICRW than SRW. Hence, we consider the chaotic maps which provide CPRN in the range (−1, 1) for generating RW algorithms. Empirical experiments reveal that step size of 20 provides a reasonable and acceptable results into coverage area of the search space by using the CCRW and ICRW algorithms for landscape analysis.

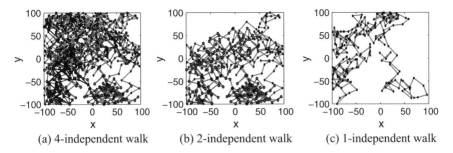

(a) 4-independent walk   (b) 2-independent walk   (c) 1-independent walk

Fig. 3.6 Visualization of different independent sample walks of 200 steps with step size 20 by CCRW algorithm

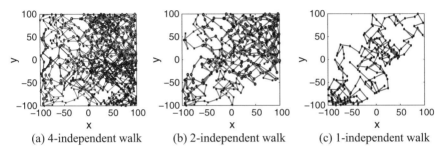

(a) 4-independent walk   (b) 2-independent walk   (c) 1-independent walk

Fig. 3.7 Visualization of different independent sample walks of 200 steps with step size 20 by ICRW algorithm

The position vectors are generated using different number of runs by the CCRW and ICRW algorithm in a 2-dimensional space as shown in Figs. 3.6 and 3.7, respectively. In contrast to Fig. 3.1, a better coverage is obtained using CRW and ICRW compared to SRW and TCRW in 2-dimension. It is worth noting that more number of a walk generated in different runs are embedded in the search space to provide coverage of the whole search space.

3.6.3 Coverage Analysis of the Walks

Deviation of a uniform sampling distribution is used to quantify the coverage area of a walk in continuous search space. A histogram is a graphical representation of the distribution of a sample of points, estimating the probability distribution of a continuous variable (quantitative variable). In our experiments, Fig. 3.8 shows the distributions of three walks of 10,000 points in a 2-dimensional space with range $[-100, 100]$. The frequencies are based on $100\,(10 \times 10)$ bins of equal size, resulting in a mean of 100 points per bin. Figure 3.8 show the distributions of the sample points obtained using SRW, CCRW and ICRW, respectively. In each case, the sample points

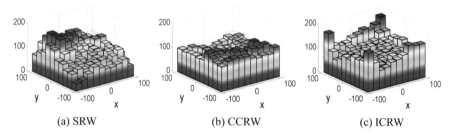

(a) SRW   (b) CCRW   (c) ICRW

Fig. 3.8 Histogram of three random walk algorithms for a sample of 10,000 points

are generated with a step size of 20 (10% of the domain). It can be observed from Fig. 3.8a that the frequencies deviate maximally from the mean of 100. Figure 3.8b shows the distribution of sample points resulting from CCRW where the frequencies deviate slightly from the mean of 100. From the histogram (Fig. 3.8), it can be seen that the distribution of sample points produced by the CCRW cover the search space more accurately than the other two random walks. The problem of clustering of points in the search space is clearly shown in the histogram of the walks as depicted in Fig. 3.8.

To test the coverage of the random walk algorithms in higher dimensions, experiments are conducted on search spaces with domain $[-100, 100]$ for dimensions (D) 1, 2, 3, 4, 5, 10, 15, 20, 25, 30, 35, 40, 45 and 50. Sample sizes or a number of points are chosen to be in the order of $10,000D$, with bin size 100.

For the four random walk algorithms, SRW, PRW, CCRW and ICRW, 30 walks were generated through independent runs of the algorithm for each dimension. Firstly, in each run, the standard deviation of the frequency in the bins are calculated. Secondly, the mean of the standard deviations through 30 independent runs are calculated. The average standard deviations from the mean frequency with standard deviations over 30 runs are reported in Table 3.1. It can be observed from Table 3.1 that a walk generated by SRW has much higher average deviations with the increase of the dimensions. A sample of points generated by a progressive random walk has average deviations ranging from approximately 9 in a single dimension to 67 in 50 dimensions. Results of the average standard deviations obtained using ICRW dominates the two walks SRW and PRW and are closer to CCRW. Points generated by the CCRW algorithm on an average have significantly smaller deviations than other walks. Results show that the points generated by the CCRW algorithm are more uniformly distributed over the search space compared to the other walks. The CCRW effectively generates CPRNs for which the step size is adaptively scaled and thus maintains a reasonable close proximity between neighboring points on the walk.

Table 3.1 Average standard deviation of the mean frequency in a bin over 30 runs

Dimensions	No. of points in walk	SRW		PRW		CCRW		ICRW	
		Mean	Std.	Mean	Std.	Mean	Std.	Mean	Std.
1	$10^4 \times 1$	20.679	2.961	9.730	0.921	8.749	0.428	9.762	0.382
2	$10^4 \times 2$	22.913	2.618	13.274	0.656	11.360	0.885	12.974	0.840
3	$10^4 \times 3$	31.667	2.319	16.378	0.667	13.555	1.057	15.370	0.877
4	$10^4 \times 4$	31.548	1.455	19.107	0.675	13.521	0.675	16.300	0.819
5	$10^4 \times 5$	31.378	1.241	21.215	0.673	15.635	0.746	19.396	0.951
10	$10^4 \times 10$	41.002	0.685	30.024	0.712	22.563	0.783	26.364	0.620
15	$10^4 \times 15$	46.712	0.421	36.662	0.699	28.423	0.874	32.780	0.559
20	$10^4 \times 20$	50.417	0.404	42.301	0.540	33.710	0.903	36.194	0.383
25	$10^4 \times 25$	52.927	0.397	47.299	0.588	34.638	0.817	41.193	0.482
30	$10^4 \times 30$	54.769	0.278	51.923	0.652	39.013	0.644	45.612	0.443
35	$10^4 \times 35$	56.078	0.304	56.194	0.885	46.316	0.944	49.499	0.458
40	$10^4 \times 40$	67.261	0.308	60.102	0.675	49.576	0.060	52.829	0.408
45	$10^4 \times 45$	70.840	0.298	63.727	0.646	57.309	0.885	58.262	0.321
50	$10^4 \times 50$	74.135	0.345	67.058	0.783	62.425	0.903	64.797	0.353

3.7   Experiments for the Benchmark Functions

To determine whether the proposed algorithm could be used as a basis of the fitness landscape analysis, we measure the ruggedness of a number of benchmark functions in different dimensions.

3.7.1   Benchmark Functions

To measure the ruggedness and deception of a function, we use all the 28 benchmark functions taken from the IEEE congress on evolutionary computing (CEC) 2013 special session on real-parameter optimization as specified in Table 3.2. A more detailed description and parameter settings of the CEC 2013 test suite can be found in [130]. In Table 3.2, f_1–f_5 are unimodal functions, f_6–f_{20} are basic multi-modal functions and f_{21}–f_{28} are composition functions. The domain of the search space is represented as $[-100, 100]^D$ (where D is dimension) for all functions. In Table 3.2, n represents number of basic functions in the composition functions.

3.7.2   Simulation Configurations

For studying the effectiveness of the proposed random walk, we measure the ruggedness and deception of the benchmark functions using the entropy based technique and FDC. The step length of the walks is set as $10,000D$ (maximum number of function evaluations are used in real-parameter optimization competitions [130]) for the benchmark functions over 30 independent runs. The step size (δ) is 20 throughout the experiments and dimensions are taken as $D = 10, 20$ and 30. For each benchmark function, the proposed chaotic random walks are used to estimate the information stability measure (ε^*) which is the largest differences in fitness values between any two successive points on the walk. ε^* also provide the smallest value of ε for which landscape structure becomes flat.

For each walk, the entropy $H(\varepsilon)$ is calculated by increasing values of ε. The behavior of $H(\varepsilon)$ for different values of ε is significant for characterizing the ruggedness of the landscape structure [110]. Therefore, a larger proportion of ε values is chosen to compare the benchmark functions with a wide range of different fitness values. We considered five values of ε obtained by increasing the value logarithmically as 10^{-2}, 10^0, 10^2, 10^4 and 10^5.

Table 3.2 The CEC 2013 real-parameter Benchmark Functions. Please refer to the details of the website http://www.ntu.edu.sg/home/EPNSugan/

Fun. no.	Name of the Benchmark Functions	Search Space
f_1	Shifted Sphere Function	$[-100, 100]^D$
f_2	Rotated High Conditioned Elliptic Function	$[-100, 100]^D$
f_3	Rotated Bent Cigar Function	$[-100, 100]^D$
f_4	Rotated Discus Function	$[-100, 100]^D$
f_5	Different Power Function	$[-100, 100]^D$
f_6	Rotated Rosenbrock s Function	$[-100, 100]^D$
f_7	Rotated Schaffers F7 Function	$[-100, 100]^D$
f_8	Rotated Ackley s Function	$[-100, 100]^D$
f_9	Rotated Weierstrass Function	$[-100, 100]^D$
f_{10}	Rotated Griewank s Function	$[-100, 100]^D$
f_{11}	Rastrigin s Function	$[-100, 100]^D$
f_{12}	Rotated Rastrigin s Function	$[-100, 100]^D$
f_{13}	Non-Continuous Rotated Rastrigin's Function	$[-100, 100]^D$
f_{14}	Schwefel s Function	$[-100, 100]^D$
f_{15}	Rotated Schwefel s Function	$[-100, 100]^D$
f_{16}	Rotated Katsuura Function	$[-100, 100]^D$
f_{17}	Lunacek Bi_Rastrigin Function	$[-100, 100]^D$
f_{18}	Rotated Lunacek Bi_Rastrigin Function	$[-100, 100]^D$
f_{19}	Expanded Griewank s Plus Rosenbrock s Function	$[-100, 100]^D$
f_{20}	Expanded Scaffer s Function	$[-100, 100]^D$
f_{21}	Composition Function 1 (n = 5, Rotated)	$[-100, 100]^D$
f_{22}	Composition Function 2 (n = 3, Unrotated)	$[-100, 100]^D$
f_{23}	Composition Function 3 (n = 3, Rotated)	$[-100, 100]^D$
f_{24}	Composition Function 4 (n = 3, Rotated)	$[-100, 100]^D$
f_{25}	Composition Function 5 (n = 3, Rotated)	$[-100, 100]^D$
f_{26}	Composition Function 6 (n = 5, Rotated)	$[-100, 100]^D$
f_{27}	Composition Function 7 (n = 5, Rotated)	$[-100, 100]^D$
f_{28}	Composition Function 8 (n = 5, Rotated)	$[-100, 100]^D$

3.7.3 Results and Discussion

3.7.3.1 Landscape Analysis for 10-Dimension

For each benchmark function, the *mean* of $H(\varepsilon)$ and *standard deviation (std.)* over 30 independent runs are calculated considering eight values of ε based on the two random walks CCRW and ICRW; and results are reported in Tables 3.3 and 3.4. Tables 3.3 and 3.4, reveal that each function has different rugged shapes at different levels of ε. No significant difference has been observed on $H(\varepsilon)$ values obtained by

Table 3.3 The entropic measure $H(\varepsilon)$ for all functions in 10-Dimensions with different values of ε using CCRW

F No.	10^{-2}		10^{0}		10^{2}		10^{4}		10^{5}	
	Mean	Std.	Mean	Std.	Mean	Std.	Mean	Std.	Mean	Std.
f_1	0.392	3.525E-4	0.394	4.645E-4	0.463	0.002	0.328	0.003	0	0
f_2	0.392	2.824E-4	0.392	2.824E-4	0.392	2.824E-4	0.392	2.903E-4	0.392	4.435E-4
f_3	0.392	3.533E-4	0.392	3.533E-4	0.392	3.533E-4	0.392	3.533E-4	0.392	3.533E-4
f_4	0.394	3.065E-4	0.394	3.065E-4	0.394	3.100E-4	0.394	3.286E-4	0.398	5.990E-4
f_5	0.383	3.837E-4	0.383	4.041E-4	0.393	9.615E-4	0.634	0.002	0.577	0.004
f_6	0.392	2.631E-4	0.395	5.254E-4	0.513	0.002	0.188	0.006	0	0
f_7	0.393	2.872E-4	0.393	3.710E-4	0.403	0.001	0.432	0.002	0.460	0.003
f_8	0.528	0.002	0.025	0.001	0	0	0	0	0	0
f_9	0.427	0.001	0.858	9.736E-4	0	0	0	0	0	0
f_{10}	0.391	2.400E-4	0.396	4.608E-4	0.575	0.002	0.006	9.184E-4	0	0
f_{11}	0.393	4.018E-4	0.427	0.001	0.738	0.003	0	0	0	0
f_{12}	0.394	4.020E-4	0.426	0.001	0.755	0.002	0.001	5.646E-4	0	0
f_{13}	0.396	2699E-4	0.428	0.001	0.766	0.003	0.001	6.317E-4	0	0
f_{14}	0.409	1.533E-4	0.417	7.134E-4	0.694	0.002	0	0	0	0
f_{15}	0.409	1.254E-4	0.417	7.432E-4	0.694	0.001	0	0	0	0

(continued)

Table 3.3 (continued)

F No.	10^{-2}		10^0		10^2		10^4		10^5	
	Mean	Std.	Mean	Std.	Mean	Std.	Mean	Std.	Mean	Std.
f_{16}	0.416	6.437E-4	0.682	0.002	0	0	0	0	0	0
f_{17}	0.395	3.231E-4	0.432	0.001	0.771	0.002	0	0	0	0
f_{18}	0.395	3.0149E-4	0.432	0.001	0.771	0.002	0	0	0	0
f_{19}	0.394	3.046E-4	0.394	3.096E-4	0.394	3.376E-4	0.416	0.002	0.492	0.003
f_{20}	0.033	0.003	1.750E-4	1.474E-4	0	0	0	0	0	0
f_{21}	0.393	3.086E-4	0.403	7.400E-4	0.723	0.002	0.040	0.004	0.029	0.003
f_{22}	0.409	1.324E-4	0.419	6.479E-4	0.727	0.002	0	0	0	0
f_{23}	0.409	1.305E-4	0.418	8.428E-4	0.732	0.002	0	0	0	0
f_{24}	0.402	4.576E-4	0.463	0.003	0.717	0.006	0	0	0	0
f_{25}	0.401	6.183E-4	0.585	0.003	0.064	0.005	0	0	0	0
f_{26}	0.394	4.024E-4	0.412	0.001	0.577	0.004	0.604	0.007	0.024	0.003
f_{27}	0.393	2.935E-4	0.408	0.001	0.774	0.001	0.019	0.002	4.414E-4	3.422E-4
f_{28}	0.396	2.725E-4	0.407	9.359E-4	0.726	0.003	0.252	0.008	0.162	0.007

Table 3.4 The entropic measure, $H(\varepsilon)$ for all functions in 10-Dimensions with different values of ε using ICRW

F No.	10^{-2}		10^{0}		10^{2}		10^{4}		10^{5}	
	Mean	Std.	Mean	Std.	Mean	Std.	Mean	Std.	Mean	Std.
f_1	0.385	3.942E-4	0.386	4.386E-4	0.453	0.002	0.340	0.003	0	0
f_2	0.381	4.068E-4	0.381	4.068E-4	0.381	4.068E-4	0.381	4.096E-4	0.382	4.743E-4
f_3	0.383	3.838E-4	0.383	3.838E-4	0.383	3.838E-4	0.383	3.838E-4	0.383	3.838E-4
f_4	0.385	3.910E-4	0.385	3.910E-4	0.385	3.910E-4	0.385	4.191E-4	0.388	7.310E-4
f_5	0.391	3.372E-4	0.391	3.651E-4	0.400	7.198E-4	0.640	0.002	0.591	0.003
f_6	0.391	3.372E-4	0.391	3.651E-4	0.400	7.198E-4	0.640	0.002	0.591	0.003
f_7	0.383	3.482E-4	0.384	3.844E-4	0.394	0.001	0.422	0.002	0.440	0.003
f_8	0.528	0.002	0.025	0.001	0	0	0	0	0	0
f_9	0.426	7.095E-4	0.856	9.405E-4	0	0	0	0	0	0
f_{10}	0.382	3.326E-4	0.386	5.925E-4	0.559	0.002	0.008	9.538E-4	0	0
f_{11}	0.390	3.453E-4	0.423	0.001	0.741	0.002	0	0	0	0
f_{12}	0.386	4.121E-4	0.417	0.001	0.741	0.002	0.002	6.281E-4	0	0
f_{13}	0.388	4.812E-4	0.419	0.002	0.754	0.002	0.003	0.001	0	0
f_{14}	0.409	1.120E-4	0.417	9.163E-4	0.694	0.002	0		0	0
f_{15}	0.409	6.970E-5	0.416	8.476E-4	0.695	0.002	0	0	0	0

(continued)

Table 3.4 (continued)

F No.	10^{-2}		10^0		10^2		10^4		10^5	
	Mean	Std.	Mean	Std.	Mean	Std.	Mean	Std.	Mean	Std.
f_{16}	0.416	7.014E-4	0.681	0.002	0	0	0	0	0	0
f_{17}	0.389	3.616E-4	0.425	0.002	0.763	0.002	0	0	0	0
f_{18}	0.390	4.157E-4	0.425	0.001	0.762	0.001	0	0	0	0
f_{19}	0.397	2.526E-4	0.397	2.563E-4	0.397	3.634E-4	0.416	0.001	0.485	0.002
f_{20}	0.033	0.003	1.899E-4	1.603E-4	0	0	0	0	0	0
f_{21}	0.385	3.447E-4	0.395	9.439E-4	0.701	0.002	0.047	0.004	0.035	0.003
f_{22}	0.409	1.765E-4	0.418	6.443E-4	0.727	0.002	0	0	0	0
f_{23}	0.409	1.238E-4	0.418	6.848E-4	0.731	0.002	0	0	0	0
f_{24}	0.394	6.581E-4	0.576	0.002	0.071	0.003	0	0	0	0
f_{25}	0.394	6.581E-4	0.576	0.002	0.074	0.003	0	0	0	0
f_{26}	0.385	5.433E-4	0.406	0.002	0.575	0.003	0.587	0.005	0.029	0.002
f_{27}	0.385	4.851E-4	0.399	0.001	0.753	0.002	0.016	0.002	3.637E-4	2.124E-4
f_{28}	0.389	2.896E-4	0.399	7.874E-4	0.707	0.003	0.258	0.006	0.172	0.006

the CCRW and ICRW for all functions. A significant value of ruggedness for each function is obtained from the maximum value of $H(\varepsilon)$ over each values of ε, because higher value of $H(\varepsilon)$ provides more variety of rugged shapes in the walk and hence the more rugged landscape. Moreover, all the $H(\varepsilon)$ values are robust because very small *standard deviation* values are achieved for each cases.

Table 3.5 summarizes *mean* and *standard deviation(std.)* of the landscape measures for each function in 10-Dimension, achieved by the CCRW and ICRW. It can be observed from Table 3.5 that entropy measure, $H(\varepsilon)$ for all the functions obtained by both the proposed chaotic random walks is approximately same or very close to each other. The highest $H(\varepsilon)$ value above 0.7 is achieved in the 17 functions out of the 28 functions. The functions f_2, f_3, f_4, f_7, f_{19} and f_{20} attained $H(\varepsilon)$ less than 0.5. However, the highest value of $H(\varepsilon)$ for the functions f_7, f_{19} and f_{20} are not quite as expected whereas all the composite functions $f_{21}-f_{22}$ attained the entropy measure at par with the highly rugged landscape structure. The behavior of $H(\varepsilon)$ for different values of ε is shown in Fig. 3.9a (for functions f_1 to f_{14}) and Fig. 3.9b (for functions f_{15} to f_{28}) based on the landscape structure achieved by CCRW. The graphs illustrate the trend of how ruggedness changes with respect to the relative values of ε that depends on the value of ε^* for each function. The value ε^* is the smallest value of ε for which landscapes become flat when all the values of the string $S(\varepsilon)$ is '0's. The graph in Fig. 3.9a shows that the functions f_1, f_6 and f_8-f_{14} converge to 0 at different points. Among the functions, f_8 becomes flat at an early stage with $\varepsilon = 10$ compare to other functions due to small information stability value (ε^*) as observed in Tables 3.3 and 3.4. Figure 3.9b reveals that most of the benchmark functions become flat at $\varepsilon = 10^5$ whereas functions f_{19}, f_{21} and f_{28} do not converge to 0 at that point. It is also expected from Table 3.5 of the information stability measures. In case of all functions, the peak i.e. the *mean* value of $H(\varepsilon)$ is attained at a particular value for ε and then decreases. Therefore, the point at which the maximum of $H(\varepsilon)$ occurs corresponds to the maximum difference in the landscape structure in the form of the maximum number of rugged elements.

In Table 3.5, r_{FD} (coefficients) value is small (close to 0) for most of the functions and even negative in f_{16}, f_{24}, f_{25} and f_{28} based on CCRW. It is also observed that FDC coefficients for each function achieved through ICRW is same with the CCRW. A small variation is observed on the functions f_8 and f_{28} between the CRW though the coefficient values are quite close to 0. r_{FD} value very close to 0 implies that fitness of the sample points on landscape structure and distances to the global minimum are highly uncorrelated. Therefore, r_{FD} quantifies a structural property of the benchmark functions which are highly multi-modal with no globally convex landscape. Moreover, a negative value of r_{FD} indicates a presence of many local optima in the landscape structure.

Table 3.5 Results of landscape measures for all functions using CCRW and ICRW in 10-Dimension

F No.	CCRW						ICRW					
	$H(\varepsilon)$		ε^*		FDC		$H(\varepsilon)$		ε^*		FDC	
	Mean	Std.	Mean	Std.	Mean	Std.	Mean	Std.	Mean	Std.	Mean	Std.
f_1	0.724	0.001	5.15E4	1.97E4	0.931	0.020	0.706	0.002	5.13E4	1.78E4	0.929	0.016
f_2	0.392	4.43E-4	1.74E10	2.09E9	0.290	0.041	0.382	4.743	1.84E10	3.72E9	0.272	0.033
f_3	0.392	3.53E-4	1.09E50	2.63E50	0.013	0.004	0.383	3.83E-4	8.50E49	1.38E50	0.013	0.004
f_4	0.397	5.99E-4	1.52E10	1.12E9	0.127	0.018	0.388	7.31E-4	1.57E10	7.01E8	0.112	0.021
f_5	0.634	0.002	7.69E5	1.67E5	0.237	0.030	0.640	0.002	7.25E5	6.41E4	0.243	0.038
f_6	0.780	0.001	4.48E4	6.99E3	0.668	0.065	0.757	0.001	5.34E4	1.00E4	0.658	0.069
f_7	0.460	0.002	5.54E21	3.97E21	0.029	0.009	0.449	0.002	2.06E22	2.24E22	0.029	0.008
f_8	0.849	7.69E-4	1.582	0.075	6.81E-4	0.004	0.850	0.001	1.606	0.072	1.14E-4	0.003
f_9	0.858	9.73E-4	14.401	1.006	0.027	0.028	0.855	9.40E-4	14.615	0.903	0.034	0.016
f_{10}	0.781	0.001	1.62E4	4.78E3	0.483	0.035	0.762	0.001	1.62E4	2.71E3	0.469	0.039
f_{11}	0.738	0.002	1.68E3	3.62E2	0.540	0.060	0.740	0.002	1.57E3	1.10E2	0.511	0.055
f_{12}	0.754	0.002	2.40E4	6.89E4	0.452	0.082	0.741	0.001	3.04E4	8.89E3	0.392	0.067
f_{13}	0.765	0.002	2.57E4	7.57E3	0.429	0.064	0.754	0.002	3.23E4	8.28E3	0.397	0.073
f_{14}	0.694	0.002	4.31E3	5.61E2	0.087	0.036	0.694	0.001	4.26E3	5.84E2	0.091	0.042

(continued)

Table 3.5 (continued)

F No.	CCRW						ICRW					
	$H(\varepsilon)$		ε^*		FDC		$H(\varepsilon)$		ε^*		FDC	
	Mean	Std.	Mean	Std.	Mean	Std.	Mean	Std.	Mean	Std.	Mean	Std.
f_{15}	0.693	0.001	4.27E3	4.61E2	0.065	0.032	0.694	0.001	4.37E3	4.65E2	0.056	0.033
f_{16}	0.681	0.002	2.17E2	7.648	−3.74E-4	0.003	0.680	0.002	2.16E2	7.339	−2.18E-4	0.003
f_{17}	0.771	0.001	1.34E3	3.22E2	0.875	0.033	0.762	0.001	1.43E3	3.38E2	0.867	0.040
f_{18}	0.771	0.001	1.49E3	2.44E2	0.871	0.038	0.762	0.001	1.50E3	3.25E2	0.851	0.042
f_{19}	0.491	0.002	7.96E7	5.54E6	0.649	0.037	0.485	0.002	8.35E7	5.84E6	0.659	0.031
f_{20}	0.055	0.004	6.050	3.21E-10	0.002	0.006	0.054	0.003	6.05E2	6.39E-9	0.002	0.008
f_{21}	0.723	0.002	4.79E14	1.34E15	0.003	0.003	0.701	0.002	2.49E16	6.00E16	0.004	0.004
f_{22}	0.727	0.001	5.08E3	4.31E2	0.094	0.061	0.726	0.001	5.28E3	5.65E2	0.075	0.062
f_{23}	0.713	0.002	5.09E3	5.59E2	0.084	0.080	0.730	0.001	5.22E3	4.59E2	0.061	0.062
f_{24}	0.717	0.005	1.69E3	2.28E2	−0.111	0.293	0.713	0.003	1.78E3	3.82E2	−0.171	0.232
f_{25}	0.771	0.002	1.44E3	9.34E1	−0.521	0.196	0.749	0.002	1.45E3	1.28E2	−0.457	0.336
f_{26}	0.604	0.006	2.60E5	2.34E4	0.149	0.088	0.586	0.004	2.71E5	2.67E4	0.115	0.080
f_{27}	0.774	0.001	1.40E5	2.60E4	0.247	0.073	0.753	0.001	1.41E5	2.56E4	0.308	0.073
f_{28}	0.725	0.003	1.71E10	1.53E10	−0.006	0.016	0.707	0.003	3.14E10	3.29E10	2.24E-5	0.023

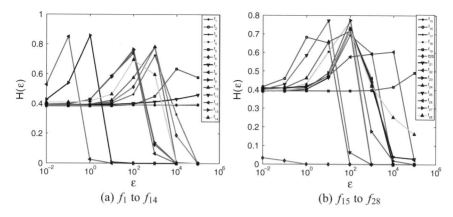

(a) f_1 to f_{14} (b) f_{15} to f_{28}

Fig. 3.9 Mean of $H(\varepsilon)$ over different values of ε for all the functions in 10-dimension using CCRW

3.7.4 Landscape Analysis for 20-Dimension

Results of *mean* and *standard deviation (std.)* of $H(\varepsilon)$ over 30 independent runs for each function at different level of ε by using CCRW and ICRW in 20-Dimension are reported in Tables 3.6 and 3.7. Tables 3.6 and 3.7 presents the rugged behaviors for each function at different levels of ε where entropic measures $H(\varepsilon)$ are approximately equal in both the proposed random walks CCRW and ICRW. It can be observed from Tables 3.3 and 3.6, a few functions converge to 0 at $\varepsilon = 10^5$ or early stages of ε as compared to the 10-Dimension cases. Table 3.8 summarizes the landscape measure results are obtained by the two proposed RWs on the test functions. *Mean* and *standard deviation (Std.)* metrics are used to compare the performance of the RWs. The entropic measure, $H(\varepsilon)$ for the functions f_2, f_3, f_4, f_7, f_{19} and f_{20} are less than 0.5 represents smooth landscape structure. However, highest value of $H(\varepsilon)$ for the functions f_7, f_{19} and f_{20} are not well as expected. The functions f_8–f_{18} have the highest $H(\varepsilon)$ value above 0.7 showing highly rugged landscape structure which is obvious from the functional properties. Out of eight composite functions, seven functions are attained $H(\varepsilon)$ value lies with in the range [0.6, 0.8) by both CCRW and ICRW which indicate the rugged landscape structure. Moreover, the rugged landscape structures are robust, justified by smaller *standard deviation* of the corresponding functions.

Figure 3.10a (for functions f_1 to f_{14}) and Fig. 3.10b (for functions f_{15} to f_{28}) presents the behavior of entropic measures for different values of ε. Figure 3.10a reveals that functions f_2–f_{14} attain approximately same initial $H(\varepsilon)$ value close to 0.4 at the $\varepsilon = 10^{-2}$. After that $H(\varepsilon)$ value increases up to certain value of ε and then start decreasing as ε increases. However, functions attain maximum $H(\varepsilon)$ at different values of ε and finally converge to 0 at $\varepsilon = 10^5$ except the functions f_2, f_3, f_4, f_5 and f_7. It is also observed from Fig. 3.10b that functions f_{15}–f_{19} and f_{21}–f_{28} attained approximately same initial $H(\varepsilon)$ value close to 0.4 at $\varepsilon = 10^{-2}$ whereas

Table 3.6 The entropic measure $H(\varepsilon)$ for all functions in 20-Dimensions with different values of ε using CCRW

F No.	10^{-2}		10^{0}		10^{2}		10^{4}		10^{5}	
	Mean	Std.	Mean	Std.	Mean	Std.	Mean	Std.	Mean	Std.
f_1	0.392	3.525E-4	0.394	4.645E-4	0.463	0.002	0.328	0.003	0	0
f_2	0.392	2.824E-4	0.392	2.824E-4	0.392	2.824E-4	0.392	2.903E-4	0.392	4.435E-4
f_3	0.393	2.231E-4	0.393	2.231E-4	0.393	2.231E-4	0.393	2.231E-4	0.393	2.231E-4
f_4	0.392	2.176E-4	0.392	2.176E-4	0.392	2.209E-4	0.393	2.265E-4	0.395	3.833E-4
f_5	0.386	2.702E-4	0.385	3.012E-4	0.394	6.319E-4	0.641	0.003	0.534	0.004
f_6	0.392	2.759E-4	0.393	3.186E-4	0.442	0.001	0.639	0.013	4.72E-5	5.017E-5
f_7	0.393	2.192E-4	0.393	2.192E-4	0.393	2.153E-4	0.394	3.021E-4	0.398	6.266E-4
f_8	0.561	0.001	4.03E-4	1.656E-4	0	0	0	0	0	0
f_9	0.421	6.812E-4	0.816	7.714E-4	0	0	0	0	0	0
f_{10}	0.392	2.179E-4	0.395	3.804E-4	0.540	0.002	0.006	5.439E-4	0	0
f_{11}	0.394	2.236E-4	0.416	8.847E-4	0.810	8.741E-4	0	0	0	0
f_{12}	0.392	2.665E-4	0.406	8.195E-4	0.764	0.002	0	0	0	0
f_{13}	0.394	2.919E-4	0.407	7.577E-4	0.764	0.002	0	0	0	0
f_{14}	0.409	1.007E-4	0.414	4.006E-4	0.640	0.001	0	0	0	0
f_{15}	0.409	9.188E-5	0.415	3.741E-4	0.640	0.002	0	0	0	0

(continued)

Table 3.6 (continued)

F No.	10^{-2}		10^{0}		10^{2}		10^{4}		10^{5}	
	Mean	Std.	Mean	Std.	Mean	Std.	Mean	Std.	Mean	Std.
f_{16}	0.418	5.161E-4	0.721	0.001	0	0	0	0	0	0
f_{17}	0.395	3.440E-4	0.418	8.017E-4	0.835	8.524E-4	0	0	0	0
f_{18}	0.395	2.725E-4	0.417	8.869E-4	0.835	6.045E-4	0	0	0	0
f_{19}	0.396	0.002	0.396	.002	0.396	0.002	0.401	0.002	0.431	9.762E-4
f_{20}	6.81E-6	2.59E-5	0	0	0	0	0	0	0	0
f_{21}	0.392	7.976E-4	0.398	8.189E-4	0.615	0.005	0.343	0.008	0.310	0.009
f_{22}	0.409	1.095E-4	0.417	5.533E-4	0.686	0.001	0	0	0	0
f_{23}	0.409	9.188E-5	0.416	5.262E-4	0.688	0.001	0	0	0	0
f_{24}	0.397	3.488E-4	0.427	9.463E-4	0.795	0.002	0.101	0.004	0	0
f_{25}	0.395	2.919E-4	0.491	0.001	0.331	0.006	0	0	0	0
f_{26}	0.395	2.280E-4	0.418	9.330E-4	0.665	0.003	0.465	0.005	0.071	0.002
f_{27}	0.392	2.462E-4	0.403	5.746E-4	0.712	0.002	0.008	0.001	9.991E-4	3.393E-4
f_{28}	0.393	1.747E-4	0.394	3.278E-4	0.453	0.002	0.535	0.004	0.525	0.003

Table 3.7 The entropic measure $H(\varepsilon)$ for all functions in 20-Dimensions with different values of ε using ICRW

F No.	10^{-2}		10^0		10^2		10^4		10^5	
	Mean	Std.	Mean	Std.	Mean	Std.	Mean	Std.	Mean	Std.
f_1	0.385	3.942E-4	0.386	4.386E-4	0.453	0.002	0.340	0.003	0	0
f_2	0.381	4.068E-4	0.381	4.068E-4	0.381	4.068E-4	0.381	4.096E-4	0.382	4.743E-4
f_3	0.384	2.628E-4	0.384	2.628E-4	0.384	2.628E-4	0.384	2.628E-4	0.384	2.628E-4
f_4	0.383	3.189E-4	0.383	3.189E-4	0.383	3.172E-4	0.383	3.524E-4	0.385	5.003E-4
f_5	0.391	2.294E-4	0.391	2.340E-4	0.399	4.518E-4	0.646	0.002	0.546	0.003
f_6	0.382	2.600E-4	0.383	3.447E-4	0.430	0.001	0.643	1.082E-4	0.003	8.668E-5
f_7	0.385	2.278E-4	0.385	2.278E-4	0.385	2.280E-4	0.386	2.750E-4	0.389	5.862E-4
f_8	0.562	0.001	4.055E-4	1.552E-4	0	0	0	0	0	0
f_9	0.420	7.458E-4	0.814	7.524E-4	0	0	0	0	0	0
f_{10}	0.382	2.740E-4	0.385	4.911E-4	0.524	9.455E-4	0.008	4.882E-4	0	0
f_{11}	0.390	2.485E-4	0.411	8.759E-4	0.798	8.201E-4	0	0	0	0
f_{12}	0.382	2.695E-4	0.386	7.548E-4	0.739	0.001	0	0	0	0
f_{13}	0.385	2.945E-4	0.398	7.461E-4	0.741	0.002	0	0	0	0
f_{14}	0.409	7.071E-5	0.415	3.485E-4	0.641	0.001	0	0	0	0
f_{15}	0.409	9.156E-5	0.415	5.093E-4	0.639	0.001	0	0	0	0

(continued)

Table 3.7 (continued)

F No.	10^{-2}		10^0		10^2		10^4		10^5	
	Mean	Std.	Mean	Std.	Mean	Std.	Mean	Std.	Mean	Std.
f_{16}	0.418	7.212E-4	0.721	0.001	0	0	0	0	0	0
f_{17}	0.388	2.582E-4	0.411	7.575E-4	0.818	7.248E-4	0	0	0	0
f_{18}	0.388	3.064E-4	0.410	7.698E-4	0.818	7.293E-4	0	0	0	0
f_{19}	0.398	1.838E-4	0.398	1.802E-4	0.398	1.972E-4	0.402	4.151E-4	0.428	8.541E-4
f_{20}	6.81E-6	2.592E-5	0	0	0	0	0	0	0	0
f_{21}	0.384	3.137E-4	0.389	6.851E-4	0.594	0.002	0.339	0.004	0.309	0.005
f_{22}	0.409	8.760E-5	0.417	4.987E-4	0.685	0.001	0	0	0	0
f_{23}	7.409	9.136E-5	0.416	5.112E-4	0.687	0.002	0	0	0	0
f_{24}	0.389	2.734E-4	0.419	6.615E-4	0.779	0.001	0	0	0	0
f_{25}	0.385	3.813E-4	0.482	0.002	0.322	0.005	0	0	0	0
f_{26}	0.387	2.618E-4	0.411	0.001	0.652	0.003	0.450	0.005	0.081	0.002
f_{27}	0.382	2.962E-4	0.393	7.573E-4	0.691	0.001	0.006	0.001	8.635E-4	3.053E-4
f_{28}	0.384	2.043E-4	0.385	3.579E-4	0.444	0.002	0.516	0.002	0.512	0.002

Table 3.8 Results of landscape measures for all test functions using CCRW and ICRW in 20-Dimension

F No.	CCRW						ICRW					
	$H(\varepsilon)$		ε^*		FDC		$H(\varepsilon)$		ε^*		FDC	
	Mean	Std.	Mean	Std.	Mean	Std.	Mean	Std.	Mean	Std.	Mean	Std.
f_1	0.662	0.001	1.10E5	3.13E4	0.835	0.028	0.646	0.001	1.12E5	3.13E4	0.830	0.034
f_2	0.392	2.86E-4	1.65E10	1.18E9	0.167	0.021	0.382	3.09E-4	1.67E10	8.78E8	0.156	0.019
f_3	0.393	2.23E-4	2.72E46	1.01E47	0.005	0.003	0.384	2.68E-4	2.25E47	9.12E47	0.005	0.003
f_4	0.395	3.83E-4	2.21E10	5.28E9	0.203	0.019	0.385	5.00E-4	2.58E10	5.46E9	0.193	0.015
f_5	0.641	0.003	9.93E5	2.89E5	0.202	0.025	0.646	0.002	9.77E5	3.18E5	0.191	0.027
f_6	0.651	0.005	1.04E5	1.15E4	0.659	0.046	0.643	0.003	1.10E5	2.18E4	0.655	0.042
f_7	0.398	6.26E-4	1.24E20	3.79E20	0.012	0.006	0.389	5.86E-4	2.04E20	3.24E20	0.012	0.005
f_8	0.871	5.72E-4	1.13	0.054	1.21E-4	0.002	0.871	6.95E-4	1.137	0.077	-6.09E-4	0.003
f_9	0.816	7.71E-4	21.648	1.254	0.004	0.008	0.814	7.52E-4	22.025	1.332	0.004	0.005
f_{10}	0.821	8.16E-4	2.11E4	6.56E3	0.4705	0.040	0.798	6.12E-4	2.10E4	7.31E3	0.448	0.043
f_{11}	0.810	8.74E-4	1.89E3	535.451	0.607	0.048	0.798	8.20E-4	1.86E3	432.295	0.613	0.046
f_{12}	0.764	0.002	9.33E3	2.03E3	0.617	0.051	0.739	0.001	1.06E4	2.01E3	0.606	0.050
f_{13}	0.764	0.002	9.53E3	1.62E3	0.604	0.049	0.741	0.002	1.18E4	2.81E3	0.604	0.051
f_{14}	0.772	0.001	8.27E3	922.982	0.044	0.032	0.772	0.001	8.24E3	910.142	0.046	0.027
f_{15}	0.773	0.001	8.01E3	733.382	0.037	0.023	0.773	0.001	7.94E3	1.12E3	0.036	0.023

(continued)

Table 3.8 (continued)

F No.	CCRW						ICRW					
	$H(\varepsilon)$		ε^*		FDC		$H(\varepsilon)$		ε^*		FDC	
	Mean	Std.	Mean	Std.	Mean	Std.	Mean	Std.	Mean	Std.	Mean	Std.
f_{16}	0.721	0.001	216.672	5.341	−1.99E-4	0.003	0.7208	0.0011	213.174	5.436	−3.89E-4	0.018
f_{17}	0.835	8.52E-4	2.96E3	549.874	0.797	0.0456	0.818	7.24E-4	3.01E3	591.422	0.766	0.053
f_{18}	0.835	6.04E-4	3.02E3	557.294	0.786	0.048	0.818	7.29E-4	3.26E3	525.324	0.778	0.036
f_{19}	0.431	9.76E-4	1.77E8	1.04E7	0.537	0.042	0.428	8.54E-4	1.85E8	1.274E7	0.549	0.048
f_{20}	7.82E-5	8.69E-5	610	0	0.006	0.011	7.46E-5	7.64E-5	610	0	0.010	0.011
f_{21}	0.617	0.006	8.63E26	3.66E27	0.009	0.004	0.612	0.002	2.08E28	8.84E28	0.009	0.004
f_{22}	0.686	0.001	9.30E3	604.558	0.055	0.034	0.685	0.001	8.95E3	705.749	0.056	0.044
f_{23}	0.688	0.001	9.06E3	829.999	0.052	0.044	0.687	0.002	9.27E3	699.597	0.048	0.038
f_{24}	0.795	0.002	1.58E4	4.16E3	0.397	0.054	0.779	0.002	1.94E4	5.04E3	0.397	0.058
f_{25}	0.768	0.002	2.31E3	489.697	0.333	0.059	0.750	0.001	2.34E3	5.888	0.346	0.054
f_{26}	0.665	0.003	3.92E5	3.96E4	0.184	0.062	0.652	0.003	4.16E5	5.00E4	0.181	0.068
f_{27}	0.712	0.002	2.90E5	7.36E4	0.346	0.074	0.691	0.001	2.96E5	7.46E4	0.372	0.066
f_{28}	0.535	0.004	7.77E16	6.61E16	0.016	0.006	0.526	0.003	1.17E17	1.07E17	0.016	0.005

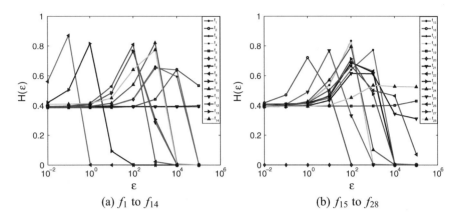

(a) f_1 to f_{14} (b) f_{15} to f_{28}

Fig. 3.10 Mean of $H(\varepsilon)$ over different values of ε for all the functions in 20-dimension using CCRW

f_{20} attained value close to 0. The value ε^* is the smallest value of ε for which the landscapes become flat. The result of information stability are shown in Tables 3.5 and 3.8 indicating that the stability measure increases with increase of dimension from 10 to 20.

It can be observed in Table 3.8 that the obtained small positive values of r_{FD} are close to 1 for the functions f_1, f_6, f_{11}–f_{13}, f_{17} and f_{18} with an exception in function f_{16} which attained small negative value very much close to 0. However, in all other functions, r_{FD} value is around 0 and does not follow any specific pattern. This fact implies that there is no correlation between the sample points on the walk and corresponding distances to the global minimum. Therefore, the r_{FD} measure provides complex landscape structures which are needle-in-a-haystack situation without any global topology regardless type of the benchmark functions.

3.7.4.1 Landscape Analysis for 30-Dimension

For each benchmark function, the *mean* and *standard deviation (Std.)* of $H(\varepsilon)$ over 30 independent runs are listed in Tables 3.9 and 3.10 by considering eight values of ε based on two random walks CCRW and ICRW. From Tables 3.9 and 3.10, it is noted that each function have different entropic measure at different level of ε. No significance difference has been observed on $H(\varepsilon)$ values obtained by the CCRW and ICRW for all the functions. No significant value of $H(\varepsilon)$ for each function is obtained compared to 20-Dimension results are given in Tables 3.6 and 3.10. Moreover, the $H(\varepsilon)$ value is robust because of very small *std.* achieved for each cases. Table 3.11 summarizes *mean* and *standard deviation* of the landscape measures for each function in 30-Dimension, achieved by the CCRW and ICRW. Table 3.11 reveals that $H(\varepsilon)$ value obtained by the CCRW and ICRW is approximately same corresponding to all the test functions. The functions f_2, f_3, f_4, f_7, f_{19} and f_{20} attained $H(\varepsilon)$ value

Table 3.9 The entropic measure $H(\varepsilon)$ for all functions in 30-Dimension with different values of ε using CCRW

F No.	10^{-2}		10^{0}		10^{2}		10^{4}		10^{5}	
	Mean	Std.	Mean	Std.	Mean	Std.	Mean	Std.	Mean	Std.
f_1	0.392	2.273E-4	0.393	2.701E-4	0.436	7.914E-4	0.700	0.001	0	0
f_2	0.390	1.822E-4	0.390	1.822E-4	0.390	1.833E-4	0.390	1.732E-4	0.391	2.216E-4
f_3	0.393	1.348E-4	0.393	1.348E-4	0.393	1.348E-4	0.393	1.348E-4	0.393	1.348E-4
f_4	0.394	3.610E-4	0.394	3.610E-4	0.394	3.521E-4	0.394	4.153E-4	0.398	4.154E-4
f_5	0.387	0.004	0.387	0.004	0.395	0.004	0.638	0.001	0.500	0.012
f_6	0.391	1.839E-4	0.392	2.342E-4	0.438	0.001	0.663	0.003	3.20E-4	9.598E-5
f_7	0.393	3.624E-4	0.393	3.624E-4	0.393	3.627E-4	0.393	4.526E-4	0.396	8.070E-4
f_8	0.585	9.211E-4	3.13E-6	1.190E-5	0	0	0	0	0	0
f_9	0.419	3.817E-4	0.782	9.355E-4	0	0	0	0	0	0
f_{10}	0.391	2.048E-4	0.393	2.594E-4	0.502	0.001	0.060	0.001	0	0
f_{11}	0.392	2.114E-4	0.402	4.286E-4	0.698	0.001	0	0	0	0
f_{12}	0.392	1.860E-4	0.408	5.499E-4	0.776	0.001	0	0	0	0
f_{13}	0.394	3.119E-4	0.409	7.584E-4	0.778	0.003	0	0	0	0
f_{14}	0.409	8.693E-5	0.414	3.298E-4	0.612	0.001	0	0	0	0

(continued)

Table 3.9 (continued)

F No.	10^{-2}		10^0		10^2		10^4		10^5	
	Mean	Std.	Mean	Std.	Mean	Std.	Mean	Std.	Mean	Std.
f_{15}	0.409	7.087E-5	0.414	2.961E-5	0.611	0.001	0	0	0	0
f_{16}	0.419	4.515E-4	0.747	7.270E-4	0	0	0	0	0	0
f_{17}	0.394	2.289E-4	0.413	5.400E-4	0.819	6.495E-4	0	0	0	0
f_{18}	0.394	2.018E-4	0.413	5.743E-4	0.891	6.878E-4	0	0	0	0
f_{19}	0.397	1.740E-4	0.397	1.740E-4	0.397	1.887E-4	0.400	4.125E-4	0.417	7.212E-4
f_{20}	5.34E-5	8.043E-5	0	0	0	0	0	0	0	0
f_{21}	0.393	4.578E-4	0.396	7.139E-4	0.549	0.004	0.437	0.006	0.403	0.008
f_{22}	0.409	7.730E-5	0.416	4.496E-4	0.656	0.001	0	0	0	0
f_{23}	0.409	7.143E-5	0.415	4.581E-4	0.658	0.001	0	0	0	0
f_{24}	0.396	2.468E-4	0.420	6.408E-4	0.819	8.727E-4	0	0	0	0
f_{25}	0.394	2.564E-4	0.473	0.001	0.379	0.009	0.004	8.656E-4	0	0
f_{26}	0.394	1.997E-4	0.415	8.048E-4	0.664	0.002	0.417	0.005	0.142	0.003
f_{27}	0.392	1.877E-4	0.401	4.790E-4	0.682	0.002	0.002	6.218E-4	2.76E-4	1.581E-4
f_{28}	0.393	5.724E-4	0.394	6.168E-4	0.431	0.004	0.531	0.004	0.546	0.002

Table 3.10 The entropic measure $H(\varepsilon)$ for all functions in 30-Dimensions with different values of ε using ICRW

F No.	10^{-2}		10^{0}		10^{2}		10^{4}		10^{5}	
	Mean	Std.	Mean	Std.	Mean	Std.	Mean	Std.	Mean	Std.
f_1	0.384	1.970E-4	0.385	2.399E-4	0.427	8.019E-4	0.684	9.776E-4	0	0
f_2	0.379	1.779E-4	0.379	1.779E-4	0.379	1.779E-4	0.380	1.725E-4	0.380	1.853E-4
f_3	0.384	2.146E-4	0.384	2.146E-4	0.384	2.146E-4	0.384	2.146E-4	0.384	2.146E-4
f_4	0.385	1.961E-4	0.385	1.961E-4	0.385	2.027E-4	0.385	2.272E-4	0.389	3.542E-4
f_5	0.390	1.943E-4	0.390	2.049E-4	0.398	4.052E-4	0.642	0.002	0.504	0.003
f_6	0.382	2.428E-4	0.382	2.719E-4	0.426	8.707E-4	0.661	0.002	5.72E-4	1.516E-4
f_7	0.384	2.324E-4	0.384	2.324E-4	0.384	2.263E-4	0.384	2.444E-4	0.387	3.726E-4
f_8	0.585	9.784E-4	1.25E-5	2.110E-5	0	0	0	0	0	0
f_9	0.418	5.073E-4	0.780	7.504E-4	0	0	0	0	0	0
f_{10}	0.381	2.922E-4	0.383	4.033E-4	0.489	0.001	0.069	0.002	0	0
f_{11}	0.390	2.329E-4	0.400	4.566E-4	0.685	0.001	0	0	0	0
f_{12}	0.383	2.372E-4	0.398	5.420E-4	0.749	0.001	0.001	3.270E-4	0	0
f_{13}	0.386	2.371E-4	0.400	5.676E-4	0.754	0.001	0.002	3.391E-4	0	0
f_{14}	0.409	7.642E-5	0.414	3.264E-4	0.612	9.236E-4	0	0	0	0
f_{15}	0.409	8.340E-5	0.414	3.494E-4	0.611	0.001	0	0	0	0

(continued)

Table 3.10 (continued)

F No.	10^{-2}		10^0		10^2		10^4		10^5	
	Mean	Std.	Mean	Std.	Mean	Std.	Mean	Std.	Mean	Std.
f_{16}	0.419	4.744E-4	0.747	0.001	0	0	0	0	0	0
f_{17}	0.388	2.308E-4	0.406	5.362E-4	0.802	7.905E-4	0	0	0	0
f_{18}	0.388	1.961E-4	0.406	6.434E-4	0.802	6.435E-4	0	0	0	0
f_{19}	0.398	1.307E-4	0.398	1.307E-4	0.398	1.457E-4	0.400	3.239E-4	0.416	5.104E-4
f_{20}	2.62E-5	4.48E-5	0	0	0	0	0	0	0	0
f_{21}	0.384	2.276E-4	0.387	3.234E-4	0.529	0.001	0.436	0.002	0.404	0.002
f_{22}	0.409	5.682E-5	0.416	3.955E-4	0.656	0.001	0	0	0	0
f_{23}	0.409	6.262E-5	0.415	3.217E-4	0.657	8.584E-4	0	0	0	0
f_{24}	0.389	2.418E-4	0.411	7.224E-4	0.800	7.820E-4	0.002	2.940E-4	0	0
f_{25}	0.385	3.018E-4	0.463	0.001	0.374	0.003	0	0	0	0
f_{26}	0.386	1.938E-4	0.407	5.980E-4	0.651	0.002	0.402	0.004	0.151	.002
f_{27}	0.382	2.679E-4	0.391	6.208E-4	0.663	0.001	0.002	5.137E-4	2.59E-4	1.721E-4
f_{28}	0.384	1.825E-4	0.385	2.436E-4	0.421	0.001	0.518	0.001	0.533	0.001

Table 3.11 Results of landscape measures for all functions using CCRW and ICRW in 30-Dimension

F No.	CCRW						ICRW					
	$H(\varepsilon)$		ε^*		FDC		$H(\varepsilon)$		ε^*		FDC	
	Mean	Std.	Mean	Std.	Mean	Std.	Mean	Std.	Mean	Std.	Mean	Std.
f_1	0.695	0.001	1.58E5	3.52E4	0.763	0.041	0.684	9.77E-4	1.82E5	2.82E4	0.763	0.036
f_2	0.391	2.21E-4	2.50E10	6.83E9	0.301	0.031	0.380	1.85E-4	2.66E10	5.08E9	0.286	0.023
f_3	0.393	1.34E-4	5.70E51	2.35E52	0.004	0.002	0.384	2.14E-4	1.41E52	5.37E52	0.004	0.003
f_4	0.398	4.15E-4	1.79E10	3.56E9	0.057	0.011	0.389	3.54E-4	1.76E10	1.31E9	0.053	0.007
f_5	0.638	0.012	1.27E6	5.37E5	0.184	0.025	0.642	0.002	1.21E6	4.28E5	0.194	0.029
f_6	0.663	0.003	1.36E5	3.31E4	0.572	0.040	0.661	0.002	1.46E5	2.49E4	0.568	0.032
f_7	0.396	8.07E-4	1.71E22	1.67E22	0.009	0.004	0.387	3.72E-4	7.92E22	1.48E23	0.009	0.004
f_8	0.862	5.98E-4	0.929	0.055	2.34E-4	0.002	0.862	5.77E-4	0.949	0.072	3.01E-5	0.002
f_9	0.782	9.35E-4	27.048	1.516	3.95E-4	0.006	0.780	7.50E-4	27.312	1.763	6.84E-4	0.004
f_{10}	0.802	7.01E-4	3.38E4	1.01E4	0.479	0.034	0.777	7.85E-4	3.24E4	9.32E3	0.461	0.047
f_{11}	0.698	0.001	5.76E3	666.36	0.427	0.045	0.685	0.001	6.46E3	1.27E3	0.437	0.047
f_{12}	0.776	0.001	2.06E4	5.35E3	0.413	0.039	0.749	0.001	2.31E4	5.16E3	0.402	0.048
f_{13}	0.778	0.003	2.19E4	5.63E3	0.438	0.055	0.754	0.001	2.58E4	8.64E3	0.403	0.042
f_{14}	0.832	8.30E-4	1.20E4	1.35E3	0.020	0.027	0.831	7.24E-4	1.22E4	1.03E3	0.033	0.019
f_{15}	0.832	7.57E-4	1.24E4	1.20E3	0.042	0.023	0.832	6.86E-4	1.28E4	1.37E3	0.040	0.026

(continued)

Table 3.11 (continued)

F No.	CCRW						ICRW					
	$H(\varepsilon)$		ε^*		FDC		$H(\varepsilon)$		ε^*		FDC	
	Mean	Std.	Mean	Std.	Mean	Std.	Mean	Std.	Mean	Std.	Mean	Std.
f_{16}	0.747	7.27E-4	214.93	4.227	5.17E-8	0.002	0.747	0.001	215.238	3.574	−2.14E-4	0.002
f_{17}	0.819	6.49E-4	4.83E3	700.85	0.718	0.033	0.802	7.90E-4	4.54E3	884.77	0.700	0.038
f_{18}	0.819	6.78E-4	4.87E3	673.10	0.699	0.046	0.802	6.43E-4	4.86E3	791.42	0.712	0.045
f_{19}	0.417	7.21E-4	2.82E8	2.22E7	0.459	0.038	0.416	5.10E-4	2.87E8	1.33E7	0.457	0.044
f_{20}	1.68E-4	1.54E-4	615	0	0.011	0.009	1.55E-4	9.81E-5	615	0	0.009	0.011
f_{21}	0.662	0.003	2.19E28	6.12E28	0.004	0.003	0.652	0.002	1.78E29	3.91E29	0.004	0.002
f_{22}	0.728	0.002	1.33E4	893.081	0.045	0.025	0.730	0.001	1.32E4	591.24	0.046	0.027
f_{23}	0.722	0.001	1.32E4	900.048	0.023	0.019	0.726	0.001	1.32E4	995.02	0.020	0.017
f_{24}	0.819	8.72E-4	3.09E4	6.95E3	0.421	0.058	0.800	7.82E-4	3.69E4	6.35E3	0.401	0.055
f_{25}	0.742	0.002	6.63E3	1.67E3	0.377	0.058	0.724	0.001	7.57E3	1.85E3	0.367	0.052
f_{26}	0.664	0.002	4.63E5	3.99E4	0.079	0.053	0.651	0.002	5.01E5	4.33E4	0.111	0.073
f_{27}	0.682	0.002	2.90E5	1.13E5	0.465	0.064	0.663	0.013	2.43E5	9.54E4	0.457	0.078
f_{28}	0.546	0.002	7.65E17	1.64E18	0.007	0.003	0.533	0.001	1.31E18	2.30E18	0.006	0.004

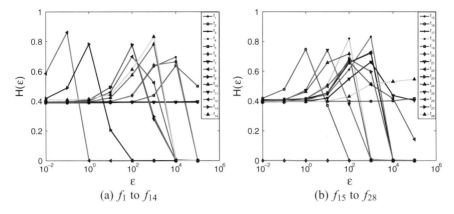

Fig. 3.11 Mean of $H(\varepsilon)$ over different values of ε for all the functions in 30-dimension

which is less than 0.5. Remaining functions have value of $H(\varepsilon)$ greater than 0.5 but in most of the functions are attained values close to 0.8, shown in Table 3.11. The $H(\varepsilon)$ value for some test functions such as f_{13}, f_{14} and f_{24} is increasing trend as dimensions increases. Similar unexpected result of $H(\varepsilon)$ is obtained in function f_{19} and f_{20} in 30-dimension as in case 10 and 20-dimension. Overall result on $H(\varepsilon)$ indicates that the test functions may be highly rugged as dimensions increase. The behavior of entropic measure for different values of ε is shown in Fig. 3.11a (for functions f_1 to f_{14}) and 3.11(b) (for functions f_{15} to f_{28}). The graphs demonstrate the trend of ruggedness changes with respect to the relative values of ε. In Fig. 3.11a, the entropic measure is same (approx. 0.4) for the test functions f_2–f_{14} at $\varepsilon = 10^{-2}$. We observed diverse nature of $H(\varepsilon)$ for different values of ε. At $\varepsilon = 10^5$, the entropy curves converges at different points indicating function landscape structure would be neutral at different value of ε. Information stability measure, ε^* for the functions f_3, f_7, f_{21} and f_{28} is increase as dimensions increase from 20 to 30. Figure 3.11b reveals that only four functions f_{19}, f_{21}, f_{27} and f_{28} do not converge to 0 whereas remaining functions f_{15}–f_{18}, f_{20}, f_{22}–f_{26} converge to 0 at $\varepsilon = 10^5$. Moreover, in Fig. 3.11b reveals that most of the benchmark functions become flat at $\varepsilon = 10^4$. It is also expected from Table 3.11 of information stability measures. A significant value of maximum $H(\varepsilon)$ for each function in 30 dimensions is analyzing the behavior of $H(\varepsilon)$ for different values of ε as shown in Table 3.9. Therefore, the point at which the maximum of $H(\varepsilon)$ occurs would correspond to the maximum difference in the landscape structure in the form of the maximum number of rugged elements.

From Table 3.11, it can be observed that r_{FD} (coefficients) values reduce with increase of dimensions. There are only two functions f_1, f_6 and f_{18} which attain the coefficient value above 0.5 and remaining functions provide small positive coefficient value close to 0 based on CCRW algorithm. It is also observed from Table 3.11 that FDC coefficients for each function achieved through ICRW are same with the CCRW. The coefficient values are very much close to 0 implies that fitness of the sample points on landscape structure and distances to the global minimum are highly uncorrelated.

Therefore, r_{FD} quantifies a structural property of the benchmark functions in 30 dimensions which are highly multi-modal with no global convex landscape.

3.7.4.2   Ruggedness Measure in Different Dimensions

We have considered ten benchmark functions from IEEE CEC 2013 suite to investigate the efficiency of the proposed chaotic random walks based on the ruggedness measure in different dimensions. The considered functions have different level of complexity such as the functions f_3, f_4 and f_5 are unimodal, f_{10}, f_{11}, f_{12}, f_{13} and f_{14} are multi-modal and f_{21}, f_{22} are composition functions. The efficiency of the proposed CCRW and ICRW algorithms are compared with SRW and PRW in various dimensions (D) as 2, 5, 10, 20, 30, 40 and 50 with respect to entropy measure ($H(\varepsilon)$). The entropy measure quantifies the ruggedness of a landscape structure for a problem. For each dimension (D) of the test functions, the *mean* of maximum $H(\varepsilon)$ considering eight values of ε over 30 independent runs are calculated and reported in Table 3.12. Table 3.12 reveals that the maximum $H(\varepsilon)$ value for the functions f_3 and f_4 is strictly below the 0.5 and for the remaining functions is strictly above the 0.6, as obtained by both the CCRW and ICRW algorithms on the continuous search space which is expected. On the basis of entropy measure, an estimation of the degree of ruggedness is defined in a fuzzy way. Therefore, when the value of $H(\varepsilon)$ less than 0.5 of a function is to be considered as not so rugged. On the other hand, when $H(\varepsilon)$ > 0.5 and tends to 1 represent increasing in ruggedness. From the experiments, the PRW algorithm does not appropriately provide the $H(\varepsilon)$ value for the functions, which is said to be rugged or not as observed in Table 3.12. For example, f_{10}, f_{11} and f_{12} have the maximum $H(\varepsilon)$ values which are less than 0.5 and tendency to 0 even dimension increases. This contradicts the nature of the functions as defined in [130]. Moreover, the values of $H(\varepsilon)$ obtained through the SRW algorithm are similar or slightly different from the values of $H(\varepsilon)$ obtained by the CCRW and ICRW algorithms with increasing dimensions.

Figure 3.12 shows the estimated ruggedness measure for the selected functions in different dimensions using different random walk algorithms. Some of the features such as unimodality, multi-modality and separability vary from dimension to dimension of a problem as mentioned in [131]. The graph for functions f_3 and f_5 in Fig. 3.12a, c exhibit that there is no change in ruggedness values as dimension increases. The maximum $H(\varepsilon)$ value for the functions f_{10} and f_{21} finally decrease as dimension increases from 20 to 50. Moreover, these represent rugged functions. The entropic value, initially decreases slightly from 0.827 to 0.695 for dimensions 2 to 10 for function f_{14} and 0.843 to 0.686 for dimensions 2 to 20 for function f_{22} by the CCRW algorithm. Thereafter, degree of ruggedness increase as dimension increases to 50. Small amount of ruggedness increases or decreases at different dimensions as observed in the rest of the functions. Therefore, proposed chaotic random walk algorithms such as CCRW and ICRW provides a flexible measure of ruggedness of the functions as changing dimensionality.

Table 3.12 Mean of maximum $H(\varepsilon)$ value over 30 runs for selected benchmark functions in different dimensions

D	RW	f_3	f_4	f_5	f_{10}	f_{11}	f_{12}	f_{13}	f_{14}	f_{21}	f_{22}
2	SRW	0.409	0.410	0.792	0.797	0.816	0.750	0.748	0.815	0.728	0.834
	PRW	0.385	0.408	0.395	0.420	0.401	0.429	0.430	0.431	0.317	0.432
	CCRW	0.396	0.398	0.607	0.679	0.726	0.740	0.743	0.827	0.722	0.843
	ICRW	0.387	0.389	0.618	0.667	0.716	0.724	0.727	0.827	0.712	0.843
5	SRW	0.408	0.409	0.786	0.706	0.752	0.772	0.772	0.744	0.805	0.776
	PRW	0.010	0.401	0.392	0.412	0.411	0.415	0.417	0.421	0.349	0.422
	CCRW	0.391	0.392	0.379	0.739	0.621	0.670	0.680	0.752	0.695	0.789
	ICRW	0.379	0.390	0.624	0.717	0.614	0.661	0.676	0.751	0.674	0.789
10	SRW	0.408	0.410	0.820	0.716	0.744	0.723	0.723	0.690	0.847	0.718
	PRW	0.014	0.406	0.393	0.414	0.109	0.315	0.308	0.417	0.020	0.419
	CCRW	0.393	0.398	0.635	0.781	0.739	0.754	0.766	0.695	0.722	0.728
	ICRW	0.383	0.389	0.640	0.762	0.740	0.742	0.755	0.694	0.701	0.727
20	SRW	0.408	0.409	0.807	0.680	0.704	0.856	0.856	0.780	0.723	0.678
	PRW	0.011	0.404	0.393	0.414	0.023	0.024	0.025	0.418	0.018	0.416
	CCRW	0.393	0.395	0.642	0.821	0.810	0.765	0.764	0.772	0.615	0.686
	ICRW	0.384	0.385	0.644	0.798	0.798	0.738	0.740	0.772	0.612	0.685

(continued)

Table 3.12 (continued)

D	RW	f_3	f_4	f_5	f_{10}	f_{11}	f_{12}	f_{13}	f_{14}	f_{21}	f_{22}
30	SRW	0.408	0.409	0.787	0.843	0.850	0.839	0.839	0.835	0.675	0.746
	PRW	0.010	0.404	0.393	0.412	0.019	0.021	0.023	0.416	0.016	0.414
	CCRW	0.392	0.397	0.639	0.801	0.697	0.775	0.777	0.831	0.662	0.727
	ICRW	0.383	0.388	0.641	0.777	0.685	0.749	0.753	0.831	0.652	0.730
40	SRW	0.408	0.409	0.806	0.811	0.827	0.854	0.854	0.859	0.669	0.774
	PRW	0.010	0.403	0.393	0.410	0.018	0.021	0.022	0.416	0.017	0.410
	CCRW	0.394	0.396	0.657	0.806	0.662	0.746	0.746	0.856	0.634	0.756
	ICRW	0.391	0.387	0.658	0.785	0.648	0.725	0.725	0.856	0.621	0.759
50	SRW	0.408	0.409	0.841	0.864	0.862	0.812	0.813	0.868	0.675	0.812
	PRW	0.010	0.405	0.394	0.407	0.012	0.022	0.023	0.410	0.017	0.409
	CCRW	0.393	0.397	0.669	0.778	0.740	0.798	0.799	0.867	0.667	0.799
	ICRW	0.383	0.388	0.665	0.754	0.727	0.773	0.776	0.867	0.646	0.801

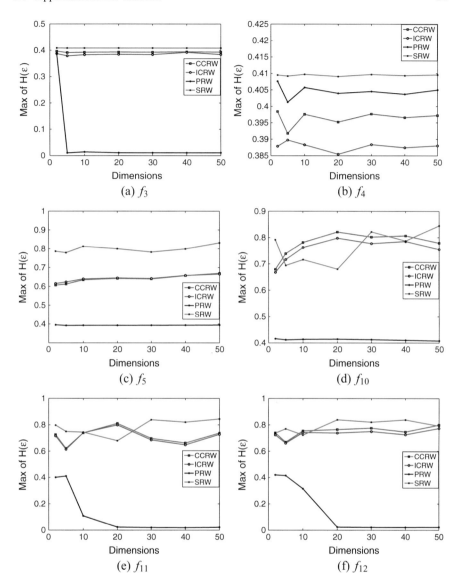

Fig. 3.12 Ruggedness characteristics for selected benchmark functions in different dimensions using four random walk algorithms

3.8 Application to PSP Problem

In order to demonstrate the efficiency of the proposed approach to real world problems, two CRWs, as well as PRW and SRW, are run on the protein structure prediction (PSP) problem.

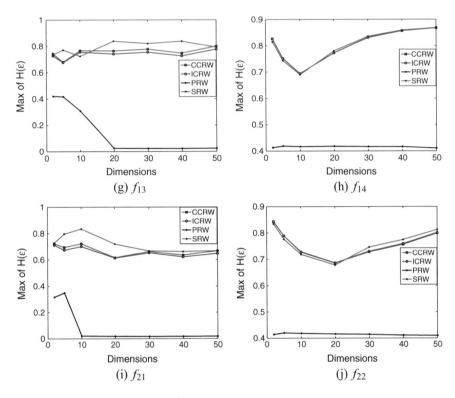

Fig. 3.12 (continued)

The PSP problem is a challenging task in computational biology for determining biological functions. Under the certain conditions, a protein sequence is folded into a stable three-dimensional structure, called native structure that consumes global minimum energy [132]. Therefore, the protein structure prediction problem transformed into a global minimization problem. The energy function is formulated on the basis of the physical model. Here, $2D$ AB off-lattice model is considered for protein energy function which is represented in terms of angles within the range $[-180°, 180°]$. So, the PSP problem based on $2D$ AB off-lattice model is a continuous optimization problem. In the experiment, four real protein sequences (1BXP, 1EDP, 1EDN and 1AGT) with different lengths are considered. The parameters of each algorithm are the same used in benchmark functions landscape analysis.

Experimental results are shown in Table 3.13, respectively. It can be observed from Table 3.13 that CCRW and ICRW can reliably extract features of the protein landscape and indicates that the landscape structures are highly rugged and contains multiple funnels. Figure 3.13 shows the characteristics of the entropy measure, $H(\varepsilon)$ over the different values of ε for all the random walk algorithms with respect to each real protein sequence.

Table 3.13 Results of landscape measures (LM) for four real protein sequences using CCRW, ICRW, PRW and SRW algorithms

Protein	LMs	SRW		PRW		CCRW		ICRW	
		Mean	Std.	Mean	Std.	Mean	Std.	Mean	Std.
1BXP	$H(\varepsilon)$	0.409	0.002	0.615	0.001	0.623	0.002	0.632	0.001
	ε^*	2.02E59	1.02E60	4.41E86	2.41E87	4.80E84	2.62E85	1.38E107	7.56E107
	r_{FD}	0.004	0.003	0.008	0.002	0.001	0.003	3.77E-4	0.003
1EDP	$H(\varepsilon)$	0.492	0.002	0.635	7.96E-4	0.637	9.32E-4	0.638	8.76E-4
	ε^*	1.11E74	6.11E74	4.87E97	2.65E98	1.79E88	9.79E88	2.16E94	1.18E95
	r_{FD}	0.003	0.002	0.001	0.002	9.22E-4	0.003	5.79E-4	0.002
1EDN	$H(\varepsilon)$	0.542	0.001	0.638	6.22E-4	0.639	7.32E-4	0.644	7.02E-4
	ε^*	4.20E70	1.80E71	2.55E95	1.39E96	5.75E88	2.41E89	6.16E103	3.37E104
	r_{FD}	0.002	0.002	6.74E-4	0.002	9.83E-4	0.002	4.72E-4	0.002
1AGT	$H(\varepsilon)$	0.623	5.31E-4	0.655	4.89E-4	0.652	5.94E-4	0.655	4.50E-4
	ε^*	4.52E76	1.85E77	2.42E99	1.32E100	5.38E96	2.94E97	3.17E107	1.73E108
	r_{FD}	0.002	0.002	5.94E-4	0.001	5.42E-4	0.001	6.97E-4	0.001

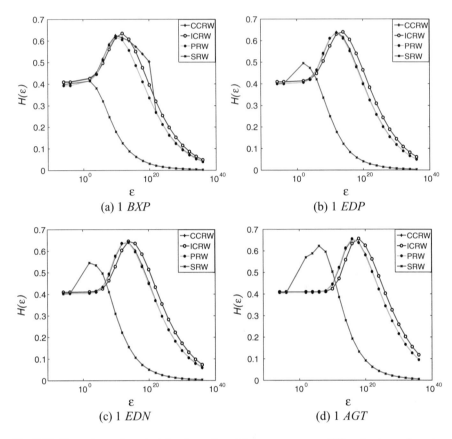

Fig. 3.13 Ruggedness characteristics for real protein sequences over different values of ε using four RW algorithms

3.9 Summary

This chapter proposed a number of random walk algorithms for continuous search spaces that could be used as the basis of techniques for characterizing the features of an optimization problem such as ruggedness, smoothness, modality, neutrality and deception. The proposed algorithm is a chaos induced random walk. The walk starts at a random position in the search space and progresses in a way with a perturbation of variable length of step size as well as the direction in each dimension of the position based on CPRNs generated by a chaotic map. It is shown that the distribution of chaotic random walk sample points over the search space is more uniform than the distribution of the simple random walk samples and hence provides a better coverage area of the entire search space. Experiments are carried on IEEE CEC 2013 benchmark test suite and four real protein instances to characterize the landscape structure on the continuous search space. Results show that the proposed landscape structure can be used to predict the features of ruggedness and deception of the problems. In addition, features indexed are also compared with the simple and progressive random walk algorithm and the results are quite favorable to our proposal.

Chapter 4
Landscape Characterization and Algorithms Selection for the PSP Problem

Abstract Fitness landscape analysis (FLA) is a technique to determine the characteristics of a problem or its structural features based on which the most appropriate algorithm is possible to recommend for solving the problem. In this chapter, we determine structural features of the protein structure prediction problem by analyzing the landscape structure. A landscape of the protein instances is generated by using the quasi-random sampling technique and city block distance. Structural features of the PSP Landscape are determined by applying various landscape measures. Numerical results indicate that the complexity of the PSP problem increases with protein sequence length. Six well-known real-coded optimization algorithms are evaluated over the same set of protein sequences and the performances are subsequently analyzed based on the structural features. Finally, we suggest the most appropriate algorithm(s) for solving the PSP problem.

4.1 Introduction

Optimization means determining the best course of action amongst the different alternatives available in a problem. It can be regarded as a process of finding the optimal value (the minimum or the maximum) of a function (usually called the objective function) under a given set of variables representing a solution to a problem. The problem is known as a continuous optimization problem when the values of the variables are real numbers. Such optimization problems are common in engineering science, technology, finance and other areas [133]. Protein structure prediction (PSP) problem is an interesting optimization problem defined over continuous domain treated as a real-world optimization problem [134]. Traditional optimization techniques such as gradient-based methods use gradient information of an objective function to determine the point where minimum or maximum value is attained in solving the problem. However, many optimization problems are not solved by using such traditional techniques due to lack of gradient information and problems are non-differential. For these type of problems, an alternative approach is used to provide the approximate solutions. Metaheuristics is the name given to such approximate solutions to any optimization problems. Metaheuristic algorithms are widely applied to real-world

© Springer International Publishing AG 2018   87
N. D. Jana et al., *A Metaheuristic Approach to Protein Structure Prediction*, Emergence, Complexity and Computation 31,
https://doi.org/10.1007/978-3-319-74775-0_4

optimization problems where variables with real values are involved in continuous spaces [135–137]. A metaheuristic framework provides many variants for solving a problem with different configurations such as choice of a search operation, numerical parameters etc. In particular, it is well-acknowledged fact that designing an algorithm configurations is essential to obtain robust and high performance of the algorithm [138–140]. Despite the development of a large number of metaheuristic algorithms, the challenge of choosing the most appropriate algorithm for solving a particular problem is a difficult task [141–145]. Researchers from the diverse fields try to deal with a new optimization problem and try to address it by using metaheuristic algorithms. The most common technique for choosing the algorithm is trial and error or tuning its configurations until a successful solution is obtained [139, 146–149]. However, it is well known that there exists no algorithm that outperforms others in solving all types of optimization problems as was proved by famous 'No-Free-Lunch' theorem [150, 151]. Therefore, the emphasis is not finding the best optimization algorithm but finding most appropriate optimization algorithm for solving a particular problem. A useful solution to this dilemma is to use Fitness Landscape Analysis(FLA) to gain an in-depth understanding of which algorithms or algorithm variants are best suited for solving a problem.

FLA is a technique to determine the characteristics of a problem or its structural features based on which the most appropriate algorithm can be recommended for solving the problem. The problem characteristics or structural features are related to the complexity of the problem and influence the performance of an algorithm [152–155] applied to solve the problem. Hence, a fundamental question arises: how to select and configure a best-suited heuristic for solving a specific problem? FLA has been applied successfully [88, 89, 156] for a better understanding of the relationship between the problem and the algorithm. Analysis of the landscape provides structural features of a problem that help us to choose the most appropriate algorithm for solving the problem. The fitness landscape is a surface in the search space that defines the fitness (objective value) for each sample point. FLA uses random, statistical or other sampling methods to generate the points in the search space. Different type of measures developed for extracting structural features [101, 103, 109, 157, 158], can be broadly divided into two groups [104, 106, 123, 159–161]. First one is information analysis measure which utilizes the fitness information, obtained from the landscaped path and the next one is statistical analysis measure based on fitness and distance of sample points in a landscape regardless of the landscaped path.

Selecting a particular algorithm based on FLA is very much appropriate to solve the PSP problem. Understanding the characteristics of the PSP problem not studied earlier and hence the selection of the algorithm is not scientific enough for solving the PSP problem. We focus on understanding the relationship between the problem and the algorithm in order to select the best-suited algorithm for solving the PSP problem thereby avoiding selection of an algorithm on an ad-hoc basis. Determining the structural features of a problem is an essential component for deciding the appropriate algorithm for solving a particular type of problem which we discuss in this chapter.

In this chapter, we determine the structural features of the PSP problem that represents problem characteristics based on which one may select the most appropriate heuristic algorithm [91]. To the best of our knowledge, it quasi-random earlier. We introduce quasi random sampling technique and a city block distance to generate the protein landscape. Information and statistical measures of FLA are used to extract the structural features of the protein landscape. Comprehensive simulations are carried out on 65 protein instances of both artificial and real protein sequences while real sequences are collected from the Protein Data Bank (PDB) on 2D and 3D AB off-lattice model. Structural features of the PSP problem consists of high rugged landscape structure with many local optima, uncorrelated sample points with respect to their fitness and presence of multi-funnels in the search landscape. Pearson correlation coefficient is calculated among the FLA measures for the 2D and 3D AB off-lattice model and it can be observed that most of the measures are uncorrelated to each other. In addition, most widely used six metaheuristic algorithms such as Genetic Algorithm (GA), Differential Evolution (DE), Particle Swarm Optimization (PSO), Artificial Bee Colony (ABC), Covariance Matrix Adaptation Evolution Strategy (CMA-ES) and Bees Algorithm (BA) are considered in our experiments due to their properties and widespread applications. The performance of the algorithms is analyzed based on the structural features on the same set of protein instances and select most appropriate algorithm(s) for solving the PSP problem. Therefore, FLA on the PSP problem provides us information for selecting the most appropriate algorithm. Without FLA it had not been possible to select the best performer algorithm to solve the PSP problem.

The chapter is organized as follows: Sect. 4.2 describes the fundamentals of fitness landscape analysis techniques. Section 4.3 presents the proposed methodology for creating protein landscape using quasi-random sampling and city block distance. Experimental settings, detail description of experimental results and discussions for landscape analysis are provided in Sect. 4.4. Algorithm performance of six widely used optimization algorithm is analyzed in Sect. 4.5. Finally, the chapter ends with a summary in Sect. 4.6.

4.2 Fitness Landscape Measures

A fitness landscape (F) is defined as a set of candidate solutions $(\mathbf{U})$ represented using a D-dimensional vector. For example, ith candidate solution $\mathbf{x}_i \in \mathbb{R}^D$, a fitness function, $f : \mathbb{R}^D \to \mathbb{R}$ and a distance metric $d : \mathbb{R}^D \to \mathbb{R}$, represented by the tuple $F = (\mathbf{x}, f, d)$. The distance d induces the definition of neighborhood $N_\phi(\mathbf{x}_i) \subset \mathbf{U}$(search space) which is generated by applying an operator ϕ to $\mathbf{x}_i$. Landscape (F) definition is also applicable to discrete and continuous optimization problems and their corresponding landscapes are treated as discrete and continuous landscape, respectively. A neighborhood in a discrete landscape is specified using Hamming distance or evolutionary operators such as crossover and mutation [88]. On the other hand, a neighborhood in a continuous landscape can be described in terms of a distance metric such as the Euclidean distance. Moreover, an infinite number of candidate solutions would be generated in continuous search space as compared to fixed number of candidate solutions in discrete search space. Hence, landscape generation in continuous search space is more difficult than a discrete domain. Various methods are employed to generate landscape structure in continuous search space such as random walk, statistical, random or other sampling techniques [118]. This section focuses on continuous FLA measures which are classified into two groups: information based measures and statistical measures due to solving the PSP problem, as discussed below.

4.2.1 Information Based Measure

Information based analysis was first proposed [109, 110] to measure ruggedness of a landscape by analyzing local neighborhood structures and interactions. Ruggedness refers to distribution of peaks and valleys counting number of local optima in a landscape structure. A rugged landscape structure is shown in Fig. 4.1. Ruggedness is estimated using a sequence of fitness values $f_1, f_2, \ldots, f_N$ obtained on a landscape path of N sample points over the search space. The sequence is represented as a time series data, defined by using a string of symbols, $S(\varepsilon) = s_1 s_2 s_3 \ldots s_{(N-1)}$, where symbol $s_i \in \{\bar{1}, 0, 1\}$ is evaluated as given in Eq. (4.1).

$$s_i(\varepsilon) = \begin{cases} \bar{1} & if \ f_i - f_{i-1} < -\varepsilon, \\ 0 & if \ |f_i - f_{i-1}| \le \varepsilon, \\ 1 & if \ f_i - f_{i-1} > \varepsilon, \end{cases} \quad (4.1)$$

Fig. 4.1 An example of a rugged landscape structure

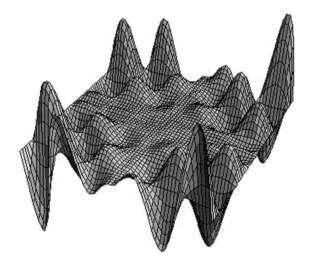

where ε is a real parameter that determines the accuracy of formation of the string $S(\varepsilon)$. A smaller value of ε provides a high sensitivity to the difference between consecutive fitness values in the landscape path. Information content, partial information content and information stability are the three measures based on information of the fitness values that quantify the ruggedness of a problem landscape.

- **Information content** refers to an entropic measure $H(\varepsilon)$, calculated as:

$$H(\varepsilon) = -\sum_{p \neq q} P_{[pq]}\log_6 P_{[pq]}, \tag{4.2}$$

where p and q are symbols of the set $\{\bar{1}, 0, 1\}$ *i.e.* $p, q \in \{\bar{1}, 0, 1\}$, and $P_{[pq]}$ calculated as:

$$P_{[pq]} = \frac{(N-1)_{[pq]}}{(N-1)}, \tag{4.3}$$

where $(N-1)_{[pq]}$ is the number of occurrences of sub-blocks pq in the string $S(\varepsilon)$. $H(\varepsilon)$ estimates variability of the shapes in the landscape path. In Eq. (4.2), all possible sub blocks pq $(p \neq q)$ of length two in the string $S(\varepsilon)$ represent shapes that are rugged. The base of the logarithmic function is 6 considering possible number of rugged shapes exist in the landscape. All possible shapes of three consecutive fitness values with class labels are shown in Table 4.1. A high

Table 4.1 Encoding and class labels of a landscape path with neighbor fitness values

Sub-block	Shape	Class label
00		Neutral
01		Rugged
0$\bar{1}$		Rugged
10		Rugged
11		Smooth
1$\bar{1}$		Rugged
$\bar{1}$0		Rugged
$\bar{1}$1		Rugged
$\bar{1}\bar{1}$		Smooth

$H(\varepsilon)$ value would provide more variety of rugged shapes in the landscape path and represents high rugged landscape structure for a problem.

- **Partial information content** estimates the modality associated with the landscape path. A new string $S'(\varepsilon)$ is constructed from the previous string of symbols $S(\varepsilon)$ represented as $S'(\varepsilon) = t_1 t_2 t_3 \ldots t_\mu$. where $t_i \neq 0$, $t_i \neq t_j$ and $\mu > 1$. We obtain the string $S'(\varepsilon)$ with length μ in the form "$\bar{1}1\bar{1}1\ldots$". The partial information content is defined as:

$$M(\varepsilon) = \frac{\mu}{(N-1)}, \tag{4.4}$$

where μ and $(N-1)$ are the length of the string $S'(\varepsilon)$ and the original string $S(\varepsilon)$ respectively. At $M(\varepsilon) = 0$, the landscape path is flat while $M(\varepsilon) = 1$ represents maximally multi-modal landscape path. Partial information content reflects the change of slopes in the landscape path which imply number of local optima in the landscape. The number of local optima is $\lfloor \frac{(N-1)M(\varepsilon)}{2} \rfloor$.

- **Information stability** is defined as the largest change in fitness between neighbors. The smallest value of ε, say, ε^* is called the information stability for which landscape becomes flat that means $S(\varepsilon^*)$ is a string of '0'. The value of ε^* is calculated as follows:

$$\varepsilon^* = min\{\varepsilon, s_i(\varepsilon) = 0\}. \tag{4.5}$$

Large value of ε^* creates high rugged landscape while small values represent smooth landscape.

A simple example for an estimation of the information based analysis follows. Assume a sequence of fitness values $\{0.1576, 0.9706, 0.9572, 0.4854, 0.8003, 0.1419\}$ belongs to the interval $[0, 1]$. We need to construct a string $S(\varepsilon) = s_1 s_2 s_3 s_4 s_5$, $(0 \leq \varepsilon \leq 1)$ to calculate $H(\varepsilon)$ and $M(\varepsilon)$. In this example, ε is set as 0.05. According to Eq. (4.1), the string $S(\varepsilon) = 10\bar{1}1\bar{1}$. The sub blocks presented in S are 10, $0\bar{1}$, $\bar{1}1$ and $1\bar{1}$ where the number of occurrences of each block is 1. Therefore, the probability of each block $P_{[01]}$, $P_{[0\bar{1}]}$, $P_{[\bar{1}1]}$, $P_{[1\bar{1}]}$ is $\frac{1}{5}$ for sequence length 5. Hence, the information content using Eq. (4.2) is

$$H(\varepsilon) = -\frac{1}{5}log_6 \frac{1}{5} - \frac{1}{5}log_6 \frac{1}{5} - \frac{1}{5}log_6 \frac{1}{5} - \frac{1}{5}log_6 \frac{1}{5}$$
$$= 0.7186$$

Next we calculate the partial information content $M(\varepsilon)$ for the string $S(\varepsilon)$ using Eq. (4.4). For $\mu = 4$, $M(\varepsilon)$ is 0.80. Therefore, the calculated $H(\varepsilon)$ and $M(\varepsilon)$ represent a landscape structure which is highly rugged and contains many local optima.

4.2.2 Statistical Measure

- **Fitness Distance Correlation (FDC)** is a statistical technique proposed by Jones et al. [104] to study the problem complexity. It determines correlation between the fitness value and respective distance to the global optimum in the search space. In the landscape, distances between sample points are evaluated using Euclidean distance. In fitness-distance analysis, the position of the global minimum is assumed to be known a priori. In real life problem like PSP, the global minimum $\mathbf{x}^*$ can be calculated using Eq. (4.6).

$$\mathbf{x}^* = \{\mathbf{x}_{i^*} | f(\mathbf{x}_{i^*}) \leq f(\mathbf{x}_i) \forall i\}, \tag{4.6}$$

where $\mathbf{x}_i$ denotes the ith point $(i = 1, 2, \ldots, N)$ in the search landscape and the corresponding fitness $f(\mathbf{x}_i) \in \mathbb{R}$. $\mathbf{x}_{i^*}$ is a sample point providing minimum fitness value among all the sample points. The fitness distance correlation coefficient r_{FD} is calculated as:

$$r_{FD} = \frac{1}{s_f s_d N} \sum_{i=1}^{N} (f_i - \bar{f})(d_i - \bar{d}), \tag{4.7}$$

where d_i is the minimum distance of $\mathbf{x}_i$ to the global minimum $\mathbf{x}^*$. $\bar{d}$ and s_d are the mean and standard deviation of the distances between the sample points to the global minimum. $\bar{f}$ and s_f are the mean and standard deviation of the fitness values over N sample points. The coefficient r_{FD} near to 1 represents global convex, single funnel landscape whereas 0 value represents needle like funnel in the landscape structure. A negative value of r_{FD} indicates 'deceiving' landscape which misleads the search processes away from the global optimum. Thus, FDC reflects the funnel structure and deception in the search landscape for a problem.
- **Autocorrelation** is a technique that measured the degree of correlation between the points on landscape path proposed in [160]. The degree of correlation depends on the difference between the fitness values of the points. Smooth landscapes are highly correlated with small difference of fitness values of the sample points. The landscape would be less correlated when the difference of fitness values is higher which implies a rugged landscape. Autocorrelation is estimated on the fitness values of neighboring sample points on the landscape path. The correlation between two points which is separated by s sample points in an N-length landscape path, calculated as:

$$\rho(s) = \frac{1}{\sigma_f^2(N-s)} \sum_{i=1}^{N-s} (f_i - \bar{f})(f_{i+s} - \bar{f}), \tag{4.8}$$

where σ_f^2 and $\bar{f}$ are the variance and mean of the fitness values of N sample points on the landscape. The correlation length τ is the distance between the uncorrelated points can be calculated as follows based on Eq. (4.8).

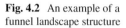

Fig. 4.2 An example of a funnel landscape structure

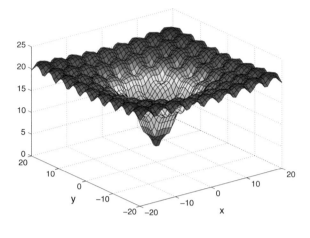

$$\tau = -\frac{1}{ln(|\rho(1)|)}, \tag{4.9}$$

where $\rho(1)$ is the correlation between two points separated by one ($s = 1$) point. The smaller value of τ indicates more rugged landscape while a higher value indicates smoother landscape.

- **Dispersion Index** [106] is a metric for estimating the presence of funnels in the landscape. A funnel in a landscape is a global basin shape that consists of clustered local optima [161]. Figure 4.2 present a funnel structure which consists huge number of local optima. Dispersion (δ) is evaluated on the sample points drawn from the set of best points in a search space. Approximate thresholds are calculated by dividing the samples into subsets with a certain percentage of best sample points of the search space. Threshold decreases as the percentage decreases. Therefore, when sample size increases with defined threshold, an increase in the dispersion indicates the presence of multiple funnels in the landscape. A positive value of δ represents the presence of funnels in the landscape and the problem is very much difficult to solve.

4.3 Energy Function Landscape Structure

FLA evaluates complexity inherent in a problem performing on a landscape path generated using sample points in the search space. FLA highly depends on the sampling method that generates the landscape-path. It is worth noting that selection of sampling method for a continuous problem is very difficult due to the existence of an infinite number of points in the search space. The goal is to generate sample points within the search space in such a way that the functional properties are achieved and well characterized for a given problem.

A simple random sampling is not free from sequential correlation on successive generations. At a time, if k random numbers at a time are used to generate the points in a D-dimensional space, the points must lie on the $(D - 1)$- dimensional planes and should not fill up the D-dimensional space. This type of sampling is sub-optimal and even inaccurate. On the other hand, Quasi-random sampling is performed where random points are carefully chosen so that the points are evenly distributed. The quasi-random sample points are efficiently generated and well spread in multidimensional search space in order to improve the asymptotic complexity of searching. Thus, the quasi-random sampling is more advantageous w.r.t every region in the search space than simple random sampling. In this study, we used Quasi-Random Sampling (QRS) and city block distance metric to generate a landscaped path in the search space.

4.3.1 Quasi-Random Sampling

Quasi-Random sampling is a deterministic alternative to simple random sampling, known as low-discrepancy sampling. How a set of points well positioned in the search space is measured by the discrepancy. A set of N number of points $\mathbf{X} = \{\mathbf{x}_1, \mathbf{x}_2, \ldots, \mathbf{x}_N\}$ where $\mathbf{x}_i = (x_{i1}, x_{i2}, \ldots, x_{iD}) \in \mathbb{R}^D$ and $B \subset \mathbf{X}$. $m(B)$ define the number of points $\mathbf{x}_i \in B$. Suppose $B_{\mathbf{x}_i}$ is the rectangular D-dimensional region, $B_{\mathbf{x}_i} = [0, x_{i1}) \times [0, x_{i2}) \times \ldots \times [0, x_{iD})$ with volume, $V_{x_1 x_2 \ldots x_D} = \prod_{j=1}^{D} x_{ij}, \forall i = 1, 2, \ldots, N$. Then the discrepancy (D_N) of the points $\mathbf{x}_1, \mathbf{x}_2, \ldots, \mathbf{x}_N$ is defined as:

$$D_N = \sup_{\mathbf{x}_i \in \mathbb{R}^D} | m(B_{\mathbf{x}_i}) - V_{x_1 x_2 \ldots x_D} |, \tag{4.10}$$

where 'sup' operator is taken over all sub-rectangles in the search space. This definition is based on the idea that for any given rectangle, the percentage of points from a uniformly distributed set which lies in the rectangle should be close to the volume of the rectangle. Therefore, smaller discrepancy, better the spacing indicate more uniformly distributed samples in the region. The low discrepancy sequences are called quasi-random. In [162], it is shown that the Sobol distribution has low discrepancy $(\mathscr{O}(\log N))$ and easy to implement compared to other distributions used in quasi-random sampling. For this reason, we used the 'Sobol' distribution for generating quasi-random sampling with an aim to generate N sample points in D-dimensional search spaces.

Figure 4.3 compares the uniformity of distributions of quasi-random sampling and pseudo-random sampling. Figure 4.3a shows a set of 500 random points generated in a 2-dimensional space within the range $[0, 1]$ using a pseudo-random number generator. It has been observed that many regions are not sampled at all and some of the regions form dense clusters by oversampling the regions, resulting in gaps in the region. Therefore, information obtained from the search space reveals the fact that the sampling is sub-optimal. Figure 4.3b shows the same number of Sobol quasi-

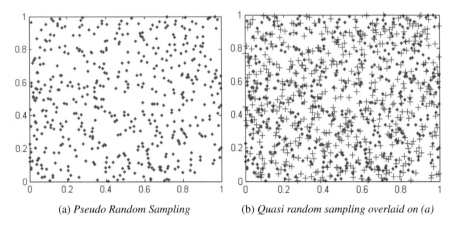

(a) *Pseudo Random Sampling*   (b) *Quasi random sampling overlaid on (a)*

Fig. 4.3 Distributions of pseudo random sampling and quasi random sampling. The quasi random sampling points do not form clusters

random samples denoted by '+' symbols embedded on the pseudo-random samples of Fig. 4.3a and fill the search spaces left by the pseudo-random samples.

4.3.2 Landscape Path Creation

A landscape path (i.e. a sequence of sample points) on protein landscape is created by using the quasi-random sampling and city block distance metric. In city block distance, an effect of a large distance in a single dimension is reduced as compared to Euclidean distance. Hence, the city block distance represents the length of walking distance. The city block distance between two points $\mathbf{x}(x_1, x_2, \ldots, x_D)$ and $\mathbf{y}(y_1, y_2, \ldots, y_D)$ with D-dimension is calculated as $\sum_{i=1}^{D} |x_i - y_i|$. The measure would be zero for identical points and high for points that are dissimilar. The basic steps of creating the proposed landscape path is described below.

Step 1 Initially, N sample points are generated using the quasi-random sampling method.

Step 2 Starting point is determined randomly from N points and marked as the current point $\mathbf{x}_1$.

Step 3 The next point $\mathbf{x}_2$ on the landscape is determined based on the minimum city block distance between $\mathbf{x}_1$ and the remaining $(N - 1)$ points.

Step 4 Repeatedly apply **Step 3** for forming the protein landscape until the landscape contains N number of sample points.

4.4 Landscape Analysis for PSP Problem

4.4.1 Experimental Settings

4.4.1.1 Protein Sequences

In order to examine the effectiveness of FLA and performance of the metaheuristic algorithms, we use two sets of protein sequences as artificial and real protein sequences as shown in Tables 4.2 and 4.3. The artificial sequences (AS) on different models have been widely used in the literature [163–165]. In the experiment, we consider 35 ASs with different lengths denoted as protein sequence length (PSL) ranging from 13 to 64 as shown in Table 4.2 where 'A' and 'B' indicates hydrophobic and hydrophilic residues respectively. ASs are clustered into five group G_{A_1}, G_{A_2}, G_{A_3}, G_{A_4} and G_{A_5} based on PSL. The G_{A_1} contains six sequence AS1-AS6 with lengths ranging from 13 to 25, G_{A_2} contains AS6-AS10 with lengths ranging from 25 to 48, G_{A_3} contains AS11-AS19 all with length 48, G_{A_4} contains AS20-AS25 with lengths ranging from 50 to 64 and G_{A_5} contains AS26-AS35 all with length 64. We consider 30 real protein sequences (RS) from the Protein Data Bank (PDB, http://www.rcsb.org/pdb/home/home.do). The information of PDB ID and PSL of RSs are given in Table 4.3. Similarly, RSs are clustered into five group G_{R_1}, G_{R_2}, G_{R_3}, G_{R_4} and G_{R_5} based on PSL. The G_{R_1} contains RS1-RS6 with lengths ranging from 13 to 21, G_{R_2} contains RS7-RS12 with lengths ranging from 21 to 26, G_{R_3} contains RS13-RS17 all with length 29, G_{R_4} contains RS18-RS23 with lengths ranging from 34 to 46 and G_{R_5} contains RS24-RS30 with lengths ranging from 54 to 98.

4.4.1.2 Simulation Strategy

Simulations for FLA measures are undertaken in MATLAB R2010b and executed on an Intel Core 2 Duo CPU with 8 GB RAM running at 2.53 GHz with Windows 7 home premium platform. Experimentally each landscape measure is evaluated using the sample points generated by the quasi-random sampling within the search space $[-180°, 180°]$ for all protein instances. Usually, the computational cost of a single search algorithm is bounded to $10,000D$ (D being the dimension of a problem) number of function evaluations (FEs) [166]. Ideally, the computational cost of FLA should be less than the algorithm cost [123]. Therefore, we set the sample size as $1,000D$ (10% of the maximum algorithm cost) as a reasonable cost at the analysis stage. To evaluate the robustness of the FLA measures, 30 independent samples are generated using different random seeds. In information analysis, information content and partial information content are estimated over different values of ε, increasing logarithmically as 10^{-6}, 10^{-4}, 10^{2}, 10^{4},..., 10^{36}. Dispersion index (δ) is calculated using the thresholds obtained by considering 5% of the total samples which are best [106]. A set of sample sizes are generated by quasi-random sampling starting with 100×2^{0} and gradually increasing as 100×2^{1},..., 100×2^{7}, respectively.

Table 4.2 Artificial protein sequences

PS No.	PSL	Sequence
AS1	13	ABBABBABABBAB
AS2	20	ABABBAABABBABABAABBABA
AS3	21	BABABBABABBABBABABBAB
AS4	23	BABBABBAAAABBABBABABBAA
AS5	24	AABBABBABBABBABBABBABBAA
AS6	25	BBABBAABBBBAABBBBAABBBBAA
AS7	34	ABBABBABABBABBABABBABABBABBABABBAB
AS8	36	BBBAABBAABBBBBAAAAAAABBAABBBBAABBABB
AS9	46	BBAAABAAABBBABAABAABBABAAAABABBAAAAABABAABBAAB
AS10	48	BBABBAABBAABBBBBAAAAAAAAAABBBBBBAABBAABBABBAAAAA
AS11	48	ABAABBAAAABAAABBAABBABAAABABAABBAABBBABBBBBBBBAA
AS12	48	AAAABAABAAAAABBABBAABBABBBBBBABBABBBBABBAABBAAABA
AS13	48	BABAABAAAAAABBABABBABAABABABBBABBAABBAABBABABBAB
AS14	48	BABAABBABAAABBAABAABBBAAAAABBABAABABABBBBBABBABAB
AS15	48	BABAABBABAAABBAABAABBBAAAAABBABAABABABBBBBABBABAB
AS16	48	AAABBBAABABAABAABAABABBBBBBBABABBABBBABBAAAAAABA
AS17	48	BABBBBABAAABABAAAABAABAABBBABABBBAAABBAABBAABBBA
AS18	48	BAABAAABAAAABBAAABBBBBBABAABBAABABBBAABABABAABBB
AS19	48	BABABBBBABABABBABAAAAAABBAAABABBABAABBABAAABBBBA
AS20	48	BAABBBBBBAABBBAAABABBABAABBABBABBAABBAAAAAAABBAA
AS21	50	AABABABABAAAABABBBABBBABBBBBABBBABBBABAAAABABABABAA
AS22	55	BABABBABABBABBABABBABABBABBABABBABBABABBABABBABBAB ABBAB
AS23	58	BABAAABAAABBAABABAABAAABABABAABBAAABBABABBBBABBABB AABBABBA
AS24	60	BBAAABAAAAAAABBBAAAAAAAAAABABBBAAAAAAAAAAAABBBBAA AAAABAABAB
AS25	64	AAAAAAAAAAAABABABBAABBAABBABBAABBAABBABBAABBAABBAB ABAAAAAAAAAAAA
AS26	64	BBAAAAABBBAABBBBBBAABBBABBBBBBBABABBBABBABBABBBBBBABB BBAABAABBABBAB
AS27	64	BBABABBABBAAABAAAABBAAABBBBABABBBABABBBABABBBBBABA BBABABBBABBABB
AS28	64	ABAABBAABABBBBBAAABAAAABBABBABAABBBABABBAAABAABABB BBBAAAAAAAABBB
AS29	64	ABBAABBABBABABBBABBBBBABBBBBBABABAAABBABABBBABABBAAB BABBABBABAAABA
AS30	64	ABBBAABBABABBBABBBABAABBBAABABAABABBABBBBABBABAAABB ABBABBAAABAAAA

(continued)

Table 4.2 (continued)

PS No.	PSL	Sequence
AS31	64	ABBAABAAAABBBBBBBAABBABBBBAABBBABBABAABABBBBAABBBBA BBBBBABBBBABAA
AS32	64	BBBBABBBABBBAAAABAABBBBBABBABAABABABBBBBABBBBBBBBB BAAAABBBBAABBA
AS33	64	BBBAAABBABABBABBAABBBABBABBAABABBBABBBBBBBABAAABAA AAABBAABBBABBA
AS34	64	ABBABBAAABBBBABABBBABAABAAAAABBBBABABABBBBABABBBAA BABBBBABBAABAB
AS35	64	BBABBABBAAABBBABABBABBABBBBBBBABBAAABBABBABBABABBBB BBAAABBBBBABAB

4.4.2 Results and Discussion

4.4.2.1 For 2D AB Off-Lattice Model

Tables 4.4 and 4.5 summarize the results of the mean and standard deviation (Std.) of the landscape measure for all the protein sequences in 2D AB off-lattice model over 30 independent runs. From Table 4.4, it is clear that the entropy measure, $H(\varepsilon)$ results are approximately same with increase in protein sequence lengths for all the artificial protein sequences while $H(\varepsilon)$ values are slowly increases with PSLs for the sequences RS1-RS27 but slightly different from RS28, RS29 and RS30 as observed in Table 4.5. For all the protein instances, entropic measure is closer to 0.7. $H(\varepsilon)$ values are robust for each of the protein instance with very small stander deviation. Entropy measure indicates that both the artificial and real protein landscapes are highly rugged. Figures 4.4 and 4.5 depict the mean results of $H(\varepsilon)$ for all the protein instances with respect to different ε values.

The entropic measure is same (approximately 0.4) for each ASs and RSs at $\varepsilon = 10^{-6}$. It is also observed the diverse nature of $H(\varepsilon)$ for different values of ε. At $\varepsilon = 10^{36}$, the entropy curves converge at different points, indicating that protein landscape may become neutral at different values of ε for different protein sequence lengths. On the other hand, the entropy curves of G_{A_3} converge at the same point due to same PSLs as shown in Fig. 4.4c and same arguments are made for the sequences of G_{A_5}. Figure 4.4f represent variations on the entropy measure for five artificial sequences taken from each group with different lengths 20, 36, 48, 55 and 64. Similar patterns have been observed in Fig. 4.5 for real protein instances. Fluctuations between the curves in Fig. 4.5f are more prominent than other sub-figures of Fig. 4.5. This situation arises due to variation in the protein sequence lengths.

Modality of the landscape path is determined by the partial information content measure, $M(\varepsilon)$. From Tables 4.4 and 4.5 it is observed that the value of $M(\varepsilon)$ is roughly constant with little fluctuations. The measure does not change significantly

Table 4.3 Real protein sequences

PS No.	PSL	PDB ID	Sequence
RS1	13	1BXP	ABBBBBBABBBAB
RS2	13	1CB3	BABBBAABBAAAB
RS3	16	1BX1	ABAABBAAAAABBABB
RS4	17	1EDP	ABABBAABBBAABBABA
RS5	18	2ZNF	ABABBAABBABAABBABA
RS6	21	1EDN	ABABBAABBBAABBABABAAB
RS7	21	1DSQ	BAAAABBAABBABABBBABBB
RS8	24	1SP7	AAAAAAAABAAABAABBAAAABBB
RS9	25	2H3S	AABBAABBBBBABBBABAABBBBBB
RS10	25	1FYG	ABAAABAABBAABBAABABABBABA
RS11	25	1T2Y	ABAAABAABBABAABAABABBAABB
RS12	26	2KPA	ABABABBBAAAABBBBABABBBBBBA
RS13	29	1ARE	BBBAABAABBABABBBAABBBBBBBBBBB
RS14	29	1K48	BAAAAAABBAAAABABBAAABABBAAABB
RS15	29	1N1U	AABBAAAABABBAAABABBAAABBBAAAA
RS16	29	1NB1	AABBAAAABABBAAABABBAAABBBAAAA
RS17	29	1PT4	AABBABAABABBAAABABBAAABBBAAAA
RS18	34	2KGU	ABAABBAABABBABAABAABABABABABAAABBB
RS19	37	1TZ4	BABBABBAABBAAABBAABBAABABBBABAABBBBBB
RS20	37	1TZ4	AAABAABAABBABABBAABBBBAABBBABAABBABBB
RS21	37	2VY5	ABBBAAAAABBBBBBAABBBAABBAABABABBAABBB
RS22	38	1AGT	AAAABABABABABAABAABBBAABBABAABBBABABAB
RS23	46	1CRN	BBAAABAAABBBBBAABAAABABAAAABBBAAAAAAAABAAABBAB
RS24	54	2K4X	BBBBABBAABABAABBBBBAABAAAAABAABBABBBBAABAABBBB BBABBBBB
RS25	60	2KAP	BBAABBABABABABBABABBBBBABAABABAABBBBBBBABBBAABAA ABBABBABBAAAAB
RS26	64	1AHO	ABBABAABBABABBBAABBABABBBBABBABABBABABBABABABAA BABBAABBABBBAAABAB
RS27	75	1HVV	BAABBABBBBBBAABABBBABBABBABABABAAAAABBBABAABBABB BABBAABBABBAABBBBBAABBBBBABBB
RS28	84	1GK4	ABABAABABBBBBABBBBABBABBBBBAABAABBBBBBAABABBBABBAB BBAABBABBBBBAABABAAAABABAABBBBBAABABBBBA
RS29	87	1PCH	ABBBAAABBBAAAABABAABAAABBABBBBBBABAAAABBBBBABABBAA BAAAAAABBABBABABABABBABBBAABAABBBAABBAAABA
RS30	98	2EWH	AABABAAAAAAABBBAAAAAABAABAABBAABABAAABBBAAAABA BAAABABBAAABAAABAAABAABBAABAAAAABAAABABBBABBAA ABAABA

Table 4.4 Results of information and statistical landscape measures for 35 artificial protein instances in 2D AB off-lattice model

PS	$H(\varepsilon)$		$M(\varepsilon)$		L_{opt}		ε^*		r_{FD}		τ		δ	
	Mean	Std.	Mean	Std.	Mean	Std.	Mean	Std.	Mean	Std.	Mean	Std.	Mean	Std.
AS1	0.65	0.006	0.66	0.012	207.85	3.08	3.41E48	9.59E48	0.014	0.021	0.387	0.012	94.30	0.41
AS2	0.65	0.005	0.66	0.008	361.80	4.60	1.04E67	3.31E67	5.56E-4	0.027	0.429	0.196	126.61	0.37
AS3	0.66	0.004	0.66	0.008	386.51	5.36	2.36E80	7.47E80	0.007	0.032	0.357	3.08E-4	130.47	0.38
AS4	0.66	0.005	0.67	0.008	434.91	7.05	4.40E63	1.38E64	0.003	0.022	0.356	0.008	137.97	0.24
AS5	0.67	0.008	0.66	0.010	452.25	8.25	1.61E68	5.08E68	0.015	0.020	0.363	0.019	141.58	0.38
AS6	0.66	0.005	0.67	0.007	481.90	8.46	1.18E80	3.74E80	0.013	0.015	0.350	0.002	144.99	0.24
AS7	0.66	0.004	0.66	0.006	696.81	8.60	1.48E74	4.68E74	−0.002	0.017	0.340	0.008	174.11	0.33
AS8	0.66	0.004	0.67	0.006	749.99	8.08	3.91E69	1.23E70	0.007	0.016	0.343	0.017	179.97	0.47
AS9	0.66	0.004	0.66	0.008	1003.60	11.59	2.62E67	8.30E67	0.003	0.017	0.337	0.019	206.57	0.43
AS10	0.66	0.003	0.67	0.006	1063.66	11.17	1.78E68	2.99E68	0.008	0.016	0.323	7.95E-4	211.69	0.35
AS11	0.66	0.004	0.66	0.008	1056.55	11.58	3.35E67	1.06E68	0.007	0.013	0.325	0.006	211.39	0.22
AS12	0.66	0.003	0.66	0.007	1055.97	13.09	1.01E70	3.21E70	−0.001	0.017	0.328	0.009	211.76	0.36
AS13	0.67	0.003	0.67	0.005	1066.03	8.67	5.64E62	1.16E63	0.001	0.013	0.328	0.012	211.71	0.37
AS14	0.66	0.004	0.67	0.004	1059.53	7.35	2.15E78	6.82E78	0.004	0.010	0.331	0.012	211.58	0.39
AS15	0.66	0.004	0.67	0.006	1058.57	11.14	3.20E66	1.00E67	0.004	0.014	0.324	0.005	211.67	0.48
AS16	0.66	0.005	0.66	0.003	1057.25	6.02	1.42E72	4.49E72	5.09E-4	0.016	0.325	0.009	211.79	0.23
AS17	0.66	0.005	0.67	0.004	1058.22	6.67	1.02E65	3.25E65	7.03E-4	0.016	0.329	0.017	211.73	0.34
AS18	0.66	0.004	0.67	0.007	1069.00	10.03	5.01E77	1.58E78	0.008	0.014	0.328	0.013	211.49	0.48
AS19	0.66	0.003	0.66	0.006	1055.67	7.42	2.22E66	7.03E66	0.005	0.014	0.327	0.010	211.58	0.31
AS20	0.66	0.004	0.67	0.006	1062.91	10.80	5.05E73	1.60E74	−0.001	0.014	0.323	0.002	211.65	0.27

(continued)

Table 4.4 (continued)

PS	$H(\varepsilon)$		$M(\varepsilon)$		L_{opt}		ε^*		r_{FD}		τ		δ	
	Mean	Std.	Mean	Std.	Mean	Std.	Mean	Std.	Mean	Std.	Mean	Std.	Mean	Std.
AS21	0.66	0.003	0.67	0.003	1115.81	10.65	3.06E72	6.89E72	−0.001	0.020	0.321	8.72E-4	216.39	0.32
AS22	0.66	0.002	0.66	0.004	1241.63	6.39	1.94E74	6.14E69	0.006	0.010	0.321	0.008	227.97	0.37
AS23	0.66	0.002	0.67	0.006	1328.88	12.82	3.07E68	8.92E68	−0.005	0.014	0.316	4.53E-4	234.62	0.25
AS24	0.66	0.003	0.69	0.003	1390.41	8.21	1.11E81	3.53E81	−1.61E-4	0.011	0.316	0.006	239.15	0.45
AS25	0.66	0.002	0.67	0.006	1496.76	12.31	3.79E85	1.19E86	−3.07E-4	0.015	0.319	0.012	247.81	0.47
AS26	0.66	0.004	0.67	0.006	1495.40	12.59	2.57E78	8.15E78	0.002	0.011	0.316	0.009	247.78	0.47
AS27	0.66	0.003	0.69	0.005	1497.96	7.93	1.29E83	4.08E83	0.004	0.015	0.312	1.22E-4	247.96	0.56
AS28	0.66	0.002	0.67	0.005	1494.33	9.57	1.66E88	5.27E88	0.005	0.015	0.316	0.008	247.56	0.53
AS29	0.66	0.004	0.66	0.004	1483.69	9.02	2.41E68	7.28E68	−0.005	0.010	0.312	0.002	247.70	0.57
AS30	0.66	0.003	0.67	0.006	1497.15	15.65	9.03E69	2.78E70	0.004	0.008	0.318	0.013	247.63	0.34
AS31	0.66	0.003	0.67	0.005	1491.82	12.24	4.30E77	1.36E78	−0.004	0.009	0.318	0.009	247.49	0.32
AS32	0.66	0.003	0.66	0.005	1488.11	8.72	2.28E79	7.22E79	0.002	0.015	0.313	0.004	247.69	0.25
AS33	0.66	0.004	0.66	0.004	1486.80	8.74	1.62E82	5.12E82	−2.51E-4	0.012	0.319	0.009	247.81	0.36
AS34	0.66	0.002	0.67	0.006	1495.09	10.75	4.94E68	7.07E68	0.004	0.011	0.315	0.008	247.71	0.44
AS35	0.66	0.004	0.67	0.006	1497.22	12.04	9.04E72	2.86E73	−0.003	0.005	0.316	0.006	247.75	0.33

Table 4.5 Results of information and statistical landscape measures for 30 real protein instances in 2D AB off-lattice model

PS	$H(\varepsilon)$		$M(\varepsilon)$		L_{opt}		ε^*		r_{FD}		τ		δ	
	Mean	Std.	Mean	Std.	Mean	Std.	Mean	Std.	Mean	Std.	Mean	Std.	Mean	Std.
RS1	0.66	0.007	0.65	0.017	208.10	5.49	2.93E65	6.84E65	0.003	0.028	0.387	0.013	94.18	0.26
RS2	0.66	0.007	0.66	0.010	207.66	3.48	8.34E55	2.63E56	0.009	0.014	0.393	0.017	94.35	0.32
RS3	0.66	0.005	0.66	0.014	274.31	6.00	1.38E69	4.39E69	0.003	0.018	0.372	0.003	109.31	0.42
RS4	0.66	0.006	0.65	0.010	292.84	6.15	1.28E60	2.74E60	3.68E-4	0.021	0.370	0.005	113.55	0.35
RS5	0.66	0.006	0.66	0.006	317.41	3.27	3.19E73	1.01E74	0.013	0.022	0.368	0.009	118.10	0.31
RS6	0.66	0.006	0.66	0.011	386.06	5.81	3.48E64	1.10E65	0.001	0.030	0.365	0.014	130.39	0.43
RS7	0.66	0.004	0.66	0.009	384.54	4.36	1.16E78	3.69E78	0.003	0.024	0.360	0.003	130.38	0.45
RS8	0.66	0.004	0.66	0.008	456.39	9.13	1.33E72	4.21E72	0.011	0.012	0.358	0.014	141.51	0.46
RS9	0.66	0.006	0.66	0.012	477.27	8.74	6.04E62	1.91E63	4.38E-4	0.020	0.361	0.017	145.37	0.39
RS10	0.66	0.007	0.66	0.009	482.19	7.93	1.01E62	3.20E62	0.004	0.024	0.355	0.011	145.47	0.60
RS11	0.66	0.005	0.66	0.009	478.17	7.23	8.72E57	2.70E58	0.007	0.031	0.354	0.0103	145.24	0.35
RS12	0.66	0.004	0.67	0.009	505.99	6.23	1.65E72	5.24E72	0.008	0.013	0.358	0.016	148.90	0.44
RS13	0.66	0.006	0.66	0.006	576.19	5.16	6.14E67	1.73E68	0.008	0.010	0.343	6.02E-4	158.76	0.28
RS14	0.65	0.004	0.66	0.009	575.23	9.08	3.06E62	6.45E62	−8.62E-4	0.024	0.346	0.007	158.78	0.29
RS15	0.66	0.004	0.66	0.008	573.91	6.57	5.65E70	1.78E71	8.48E-5	0.018	0.347	0.008	158.84	0.38
RS16	0.66	0.006	0.66	0.008	576.87	5.18	1.18E66	3.74E66	0.008	0.013	0.347	0.011	158.74	0.39
RS17	0.66	0.002	0.67	0.009	579.18	6.42	1.31E72	4.09E72	0.004	0.015	0.343	1.45E-5	158.93	0.37
RS18	0.66	0.005	0.66	0.008	698.58	7.64	3.23E81	1.02E82	0.006	0.019	0.340	0.008	174.13	0.52
RS19	0.66	0.005	0.66	0.006	768.45	5.95	2.76E66	8.50E66	0.006	0.011	0.338	0.010	182.79	0.31
RS20	0.66	0.006	0.66	0.007	774.79	8.94	4.55E76	1.43E77	0.014	0.014	0.336	0.009	182.81	0.29
RS21	0.66	0.004	0.66	0.003	775.75	5.10	5.55E63	1.74E64	0.007	0.015	0.338	0.011	182.65	0.48

(continued)

Table 4.5 (continued)

PS	$H(\varepsilon)$		$M(\varepsilon)$		L_{opt}		ε^*		r_{FD}		τ		δ	
	Mean	Std.	Mean	Std.	Mean	Std.	Mean	Std.	Mean	Std.	Mean	Std.	Mean	Std.
RS22	0.66	0.003	0.67	0.006	804.81	9.77	1.82E67	5.75E67	0.007	0.011	0.334	0.008	185.56	0.31
RS23	0.66	0.003	0.67	0.007	1008.86	12.44	2.75E77	8.71E77	0.011	0.011	0.329	0.008	206.72	0.39
RS24	0.66	0.004	0.67	0.004	1228.07	6.29	4.68E69	1.46E70	0.004	0.019	0.321	0.008	225.96	0.36
RS25	0.65	0.005	0.66	0.007	1377.89	12.31	1.86E72	5.88E72	0.006	0.007	0.318	0.012	239.13	0.47
RS26	0.66	0.003	0.67	0.004	1489.52	4.94	3.36E84	1.06E85	0.003	0.015	0.317	0.011	247.95	0.23
RS27	0.66	0.003	0.67	0.005	1799.63	11.87	2.11E69	6.67E69	0.002	0.005	0.311	0.008	269.82	0.36
RS28	0.66	0.002	0.67	0.004	2060.92	16.02	1.29E75	4.08E75	0.002	0.012	0.311	0.009	286.50	0.44
RS29	0.66	0.002	0.67	0.004	2133.35	12.65	4.62E73	9.92E73	0.010	0.011	0.304	0.003	292.00	0.28
RS30	0.66	0.003	0.67	0.005	2453.91	17.18	9.71E77	3.07E78	0.004	0.008	0.300	0.007	311.11	0.49

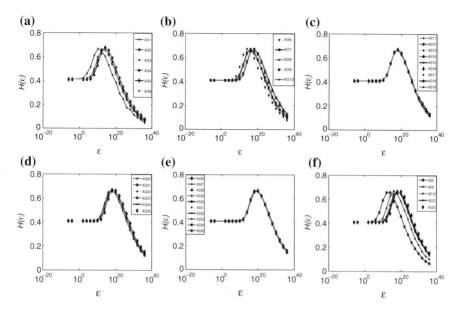

Fig. 4.4 Mean entropy, $H(\varepsilon)$ for ASs in 2D model: **a** AS1 - AS6 $\in G_{A_1}$, **b** AS6 - AS10 $\in G_{A_2}$, **c** AS11 - AS19 $\in G_{A_3}$, **d** AS20 - AS25 $\in G_{A_4}$, **e** AS26 - AS35 $\in G_{A_5}$ and **f** AS2, AS8, AS18, AS22 and AS35 taken from each group

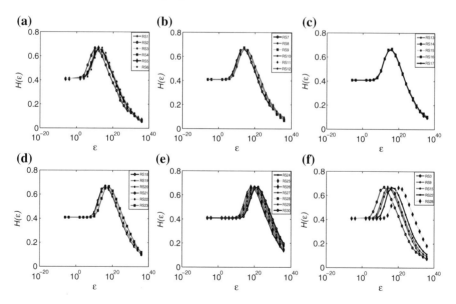

Fig. 4.5 Mean entropy, $H(\varepsilon)$ for RSs in 2D model: **a** RS1 - RS6 $\in G_{R_1}$, **b** RS7 - RS12 $\in G_{R_2}$, **c** RS13 - RS17 $\in G_{R_3}$, **d** RS18 - RS23 $\in G_{R_4}$, **e** RS24 - RS30 $\in G_{R_5}$ and **a** RS3, RS9, RS15, RS22 and RS28 taken from each group

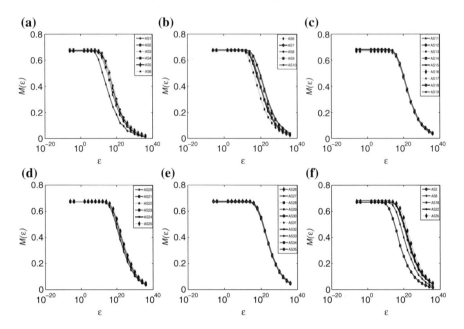

Fig. 4.6 Mean partial information content measure, $M(\varepsilon)$ curves for artificial protein sequences in 2D AB off-lattice model: **a** AS1 - AS6 $\in G_{A_1}$, **b** AS6 - AS10 $\in G_{A_2}$, **c** AS11 - AS19 $\in G_{A_3}$, **d** AS20 - AS25 $\in G_{A_4}$, **e** AS26 - AS35 $\in G_{A_5}$ and **f** AS2, AS8, AS18, AS22 and AS35 taken from each group

from sequence to sequence and the value is close to 0.7 indicates the existence of multi-modality on the protein landscape structure. The characteristics of the landscape modality for both the artificial and real protein sequences are shown in Figs. 4.6 and 4.7.

Contrary to $H(\varepsilon)$, $M(\varepsilon)$ attained highest value at $\varepsilon = 10^{-6}$ and then decreases as ε increases for all the protein instances. Moreover, the curves do not converge to a single point due to the different level of neutrality except the same length of the protein sequences. $M(\varepsilon)$ indicates that both the artificial and real protein sequences have multi-modal landscape structure. Consequently, the number of local optima (L_{opt}) are significantly increases with lengths of the protein sequences, shown in Tables 4.4 and 4.5. Information stability (ε^*) provides the smallest value of ε for which the landscape structure becomes flat or neutral. From the results, it is clear that ε^* increases with the sequence length. However, fluctuations are repeated on information stability for the artificial and real protein sequences, as observed in Tables 4.4 and 4.5. Different values of ε^* reveal that the nature of flatness of the landscape can be different from sequence to sequence. Thus, the information stability measure highly depends on the sampling methods and fitness values.

In Tables 4.4 and 4.5, r_{FD} (coefficient) values are typically very small for all the artificial and real protein sequences. However, an exception of this trend occurs in the

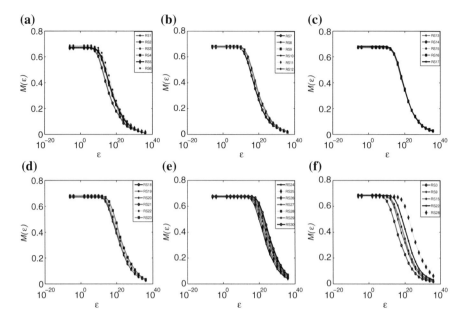

Fig. 4.7 Mean partial information content measure, $M(\varepsilon)$ curves for real protein sequences in 2D AB off-lattice model: **a** RS1 - RS6 $\in G_{R_1}$, **b** RS7 - RS12 $\in G_{R_2}$, **c** RS13 - RS17 $\in G_{R_3}$, **d** RS18 - RS23 $\in G_{R_4}$, **e** RS24 - RS30 $\in G_{R_5}$ and **a** RS3, RS9, RS15, RS22 and RS28 taken from each group

case of AS7, AS12, AS20, AS23, AS24, AS25, AS29, AS31, AS33, AS35 and RS14 protein sequences. Coefficient values are in irregular fluctuations and relatively closer to 0 as protein sequence lengths increase and indicate that fitness of the sampling points and distances to the global minimum are highly uncorrelated. Therefore, r_{FD} quantifies a structural feature of the protein landscapes which are highly multi-modal with no global convex landscape. Moreover, the negative value of r_{FD} indicates that many local optima exist on the landscape structure.

Fitness distance scatter plots are depicted in Figs. 4.8 and 4.9 for the protein sequences in 2D AB off-lattice model. Figure 4.8a–e represent the scatter plots for the sequences AS4, AS7, AS15, AS24 and AS35 taken from each group with various sequence lengths. The sequences such as AS2, AS8, AS18, AS22 and AS35 are embedded in Fig. 4.8f and indicates the variation on scatter plots among the different length of protein sequences. Similarly, scatter plots for the RSs such as RS2, RS9, RS15, RS23 and RS30 are shown in Fig. 4.9a–d and 4.9a–e respectively. Figure 4.9f reveals similar observations for the real protein sequences as in artificial protein sequences. Scatter plot provides information about the distribution of the sample points where the fitness of a sample point is plotted with respect to its distance to global minimum. In the scatter plots, different shapes are formed by the sample distribution. In Fig. 4.8 the scatter plots of sample points are scattered in outer layer of the concentrated area of the samples. Furthermore, many samples are far away from

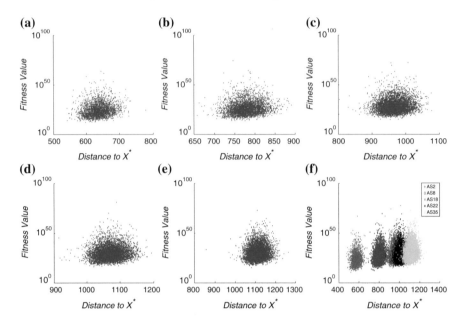

Fig. 4.8 Scatter plots of fitness versus distance to the global minimum (X^*) for artificial protein sequence in 2D AB off-lattice model: **a** AS4 with length 23, **b** AS7 with lenght 34, **c** AS21 with lenght 50, **d** AS24 with lenght 60, **e** AS34 with lenght 64 and **f** AS2, AS8, AS18, AS22 and AS35 taken from each group

the minimum distance that have lower fitness than samples close to the global mini-mum, indicating a deceptive problem. Fitness difference affects the value and range of the distances, evident for the fluctuation between fitness and global minimum, $\mathbf{x}^*$.

The results of the correlation length (τ) are reported in Tables 4.4 and 4.5. In the case of both protein sequences, τ values are close to 0 and decreases with the length of the sequence. It indicates that the degree of non-linear correlation increases with sequence length. Therefore, ruggedness of the protein landscapes are highly dependent on the length of the protein sequence. Autocorrelation values are scaled up by multiplying 10^6 for clear visualization in correlogram curves. Correlograms for artificial and real protein sequences are shown in Figs. 4.10 and 4.11. Autocorrelation values for the sequences AS1, AS4 and AS6 are closer to 10 over the different lag values. It indicates that for a set of sample points, energy value of the sequences AS1, AS4 and AS6 are quite dissimilar as compared to the other artificial protein sequences. Therefore, the ruggedness of artificial protein landscape increases with the lengths of the protein sequences. Figure 4.11 reveals the similar observations for the real protein sequences.

Dispersion metric of FLA, estimates the presence of funnels in the landscape structure and denoted by dispersion index, δ. The mean of bound-normalized δ value for 30 independent runs by increasing the sample size for both the protein sequences are shown in Tables 4.4 and 4.5. We obtain δ value for all cases indicating a sig-nificant increase in dispersion with length. High value of δ indicates the existence of multi-funnels in the landscape structure and best solution is far away from the

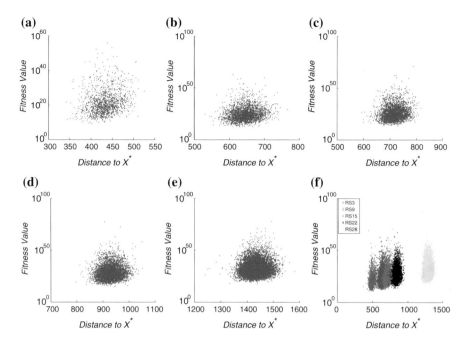

Fig. 4.9 Scatter plots of fitness versus distance to the global minimum ($\mathbf{x}^*$) for real protein sequences in 2D AB off-lattice model: **a** RS1 with lenght 13, **b** RS10 with lenght 25, **c** RS15 with lenght 29, **d** RS23 with lenght 46, **e** RS30 with lenght 98 and **f** RS3, RS9, RS15, RS22 and RS28 taken from each group

possible solutions. Hence, a significant change can be occurred in basin shape consisting of clustered local optima in the protein landscape as protein lengths increases. Figures 4.12 and 4.13 shows the dispersion characteristics of all the ASs and RSs with the increase of sample size and indicates average dispersion is a function of sample size for all the protein sequences. In Fig. 4.12a, dispersion plots are overlapped except AS1 due to relatively closer lengths of the real protein sequences. From Fig. 4.12c, it is observed that dispersion curves of the ASs $\in G_{A_3}$ are fully overlapped in a curve due to same length of the sequences which is justified by the results of dispersion in Table 4.4. Similar observations are made on ASs of G_{A_5} and are shown in Fig. 4.12e. This situation does not occur in Fig. 4.12f for variation in protein sequence length.

We apply Pearson's correlation ($\rho_{x,y}$) test for investigating the relationship between any two landscape measures, representing the strength of association between the measures are shown in Tables 4.6 and 4.7 for the artificial and real protein sequences. Boldfaced values represent a strong correlation ($\rho_{x,y} \geq 0.7$). In Table 4.6, focusing on the results of L_{opt} and δ, we observed a strong correlation ($\rho_{x,y} \approx 1$) between them. However, $H(\varepsilon)$, $M(\varepsilon)$, (ε^*), r_{FD} and τ can not be correlated with any other measures. On the other hand, τ has strong negative correlation with $H(\varepsilon)$, $M(\varepsilon)$, L_{opt}, ε^*, and δ whereas weak correlation with r_{FD}. Correlation matrix provides conformation about maximum number of landscape measures can be

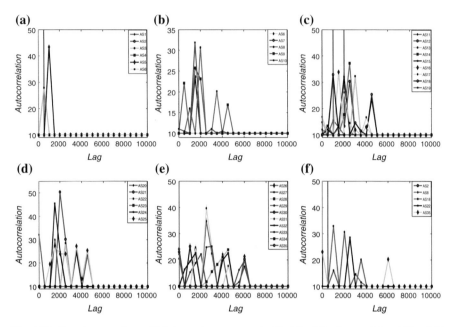

Fig. 4.10 Correlograms for artificial protein sequences in 2D AB off-lattice model: **a** AS1 - AS6 $\in G_{A_1}$, **b** AS6 - AS10 $\in G_{A_2}$, **c** AS11 - AS19 $\in G_{A_3}$, **d** AS20 - AS25 $\in G_{A_4}$, **e** AS26 - AS35 $\in G_{A_5}$ and **f** AS2, AS8, AS18, AS22 and AS35 taken from each group

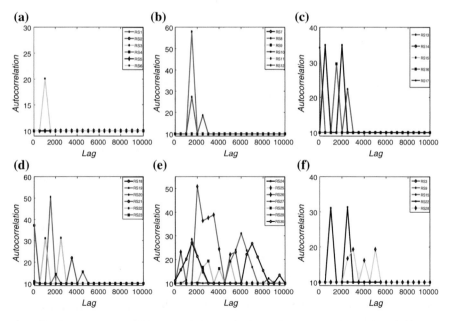

Fig. 4.11 Correlograms for real protein sequences in 2D AB off-lattice model: **a** RS1 - RS6 $\in G_{R_1}$, **b** RS7 - RS12 $\in G_{R_2}$, **c** RS13 - RS17 $\in G_{R_3}$, **d** RS18 - RS23 $\in G_{R_4}$, **e** RS24 - RS30 $\in G_{R_5}$ and **a** RS3, RS9, RS15, RS22 and RS28 taken from each group

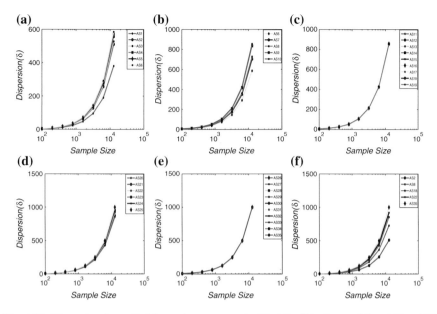

Fig. 4.12 Dispersion for artificial protein sequences in 2D AB off-lattice model: **a** AS1 - AS6 $\in G_{A_1}$, **b** AS6 - AS10 $\in G_{A_2}$, **c** AS11 - AS19 $\in G_{A_3}$, **d** AS20 - AS25 $\in G_{A_4}$, **e** AS26 - AS35 $\in G_{A_5}$ and **f** AS2, AS8, AS18, AS22 and AS35 taken from each group

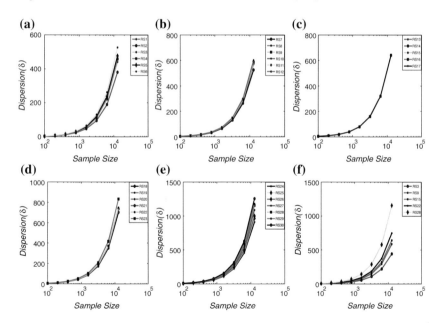

Fig. 4.13 Dispersion for real protein sequences in 2D AB off-lattice model: **a** RS1 - RS6 $\in G_{R_1}$, **b** RS7 - RS12 $\in G_{R_2}$, **c** RS13 - RS17 $\in G_{R_3}$, **d** RS18 - RS23 $\in G_{R_4}$, **e** RS24 - RS30 $\in G_{R_5}$ and **a** RS3, RS9, RS15, RS22 and RS28 taken from each group

Table 4.6 Pearson correlation coefficients ($\rho_{x,y}$) between landscape measures for all artificial protein sequences in 2D AB off-lattice model

Measure	$M(\varepsilon)$	L_{opt}	ε^*	r_{FD}	τ	δ
$H(\varepsilon)$	0.393	0.069	-0.058	0.044	-0.392	0179
$M(\varepsilon)$		0.653	0.094	-0.233	-0.735	0.680
L_{opt}			0.182	-0.550	-0.853	**0.990**
ε^*				0.087	-0.116	0.160
r_{FD}					0.398	-0.565
τ						-0.883

Table 4.7 Pearson correlation coefficients ($\rho_{x,y}$) between landscape measures for all real protein sequences in 2D AB off-lattice model

Measure	$M(\varepsilon)$	L_{opt}	ε^*	r_{FD}	τ	δ
$H(\varepsilon)$	**0.999**	0.468	0.070	0.472	**0.983**	**0.759**
$M(\varepsilon)$		0.645	0.123	0.062	-0.768	**0.719**
L_{opt}			0.207	-0.043	-0.914	**0.983**
ε^*				-0.135	-0.218	0.228
r_{FD}					-0.014	-0.017
τ						-0.969

required to extract the structural features for ASs in 2D AB off-lattice model. From Table 4.7, it is clear that $H(\varepsilon)$ is strongly correlated ($\rho_{x,y} \approx 1$) with $M(\varepsilon)$, τ and δ. Moreover, δ is strongly correlated with $H(\varepsilon)$, $M(\varepsilon)$, and L_{opt}. r_{FD} is not correlated with any other measures. In τ, it has strong negative correlation with $M(\varepsilon)$, L_{opt}, ε^*, r_{FD} and δ whereas $H(\varepsilon)$ and $M(\varepsilon)$ have weak correlation with ε^* and r_{FD}.

4.4.2.2   For 3D AB Off-Lattice Model

The *mean* and *standard deviation (std.)* results of the landscape measures for the artificial and real protein sequences in 3D AB off-lattice model are summarized in Tables 4.8 and 4.9.

The information content, $H(\varepsilon)$ for the artificial protein sequences AS1-AS6, AS8-AS21 are more or less identical and closer to 0.7 shown in Table 4.8. The sequences AS7, AS22-AS35 attained smaller value than the other artificial protein sequences and the difference is very small, approximately 0.01. No significant changes made on $H(\varepsilon)$ values are observed for artificial protein sequences as sequence lengths increases and the value is near to 0.7. Therefore, it can be inferred that the artificial protein sequence landscapes of the 3D off-lattice model are highly rugged. Moreover, the rugged landscape structures are robust, justified by smaller std. of the corresponding sequences. On 17 out of the 30 real protein sequences, $H(\varepsilon)$ values are reached above 0.670 with smaller std. deviation. These values do not fluctuate significantly

Table 4.8 Results of information and statistical measures for 35 artificial protein landscapes in 3D AB off-lattice model

PS	$H(\varepsilon)$		$M(\varepsilon)$		L_{opt}		ε^*		r_{FD}		τ		δ	
	Mean	Std.	Mean	Std.	Mean	Std.	Mean	Std.	Mean	Std.	Mean	Std.	Mean	Std.
AS1	0.67	0.007	0.66	0.012	140.64	2.11	4.18E43	1.32E44	−0.002	0.030	0.388	0.019	147.28	0.34
AS2	0.67	0.007	0.66	0.012	240.03	4.81	6.64E43	2.09E44	0.009	0.020	0.367	0.014	190.95	0.41
AS3	0.68	0.008	0.66	0.011	254.89	3.16	4.89E53	1.53E54	0.010	0.018	0.360	0.007	196.48	0.35
AS4	0.68	0.009	0.66	0.007	283.82	3.12	7.56E39	2.18E40	−0.004	0.027	0.359	0.011	207.02	0.32
AS5	0.68	0.009	0.67	0.011	302.36	4.35	7.41E43	1.55E44	0.007	0.016	0.364	0.016	211.85	0.46
AS6	0.67	0.004	0.67	0.010	315.97	4.41	5.09E46	1.61E47	2.44E-4	0.020	0.355	0.011	216.75	0.46
AS7	0.67	0.007	0.67	0.006	453.37	3.88	1.01E44	2.11E44	−0.004	0.016	0.342	0.013	256.46	0.33
AS8	0.67	0.004	0.66	0.007	484.69	4.56	2.66E50	8.43E50	0.006	0.021	0.336	0.006	264.74	0.38
AS9	0.67	0.004	0.67	0.006	648.17	6.74	1.89E46	5.99E46	−0.001	0.020	0.330	0.009	301.67	0.38
AS10	0.68	0.003	0.66	0.005	675.27	5.09	1.11E56	3.53E56	0.008	0.011	0.323	8.64E-4	308.91	0.38
AS11	0.67	0.005	0.66	0.006	678.65	8.61	6.76E51	2.13E52	−0.008	0.014	0.323	0.001	308.48	0.43
AS12	0.67	0.005	0.66	0.009	676.35	9.28	2.56E56	7.94E56	−4.07E-4	0.015	0.335	0.018	308.71	0.29
AS13	0.67	0.004	0.66	0.005	680.08	4.18	8.83E47	2.39E48	0.008	0.010	0.327	0.009	308.45	0.38
AS14	0.68	0.005	0.67	0.004	675.59	5.192	9.16E48	2.89E49	1.30E-4	0.016	0.329	0.010	308.67	0.41
AS15	0.67	0.004	0.66	0.005	677.07	4.88	6.72E43	1.41E44	0.003	0.015	0.325	0.004	308.71	0.29
AS16	0.68	0.006	0.67	0.004	678.18	4.27	1.25E46	2.80E46	−6.29E-4	0.017	0.325	0.009	308.29	0.35
AS17	0.68	0.007	0.67	0.009	680.79	7.75	4.82E58	1.52E59	−0.004	0.015	0.323	3.71E-4	308.52	0.41
AS18	0.68	0.004	0.66	0.010	676.52	10.70	5.95E52	1.88E53	−0.002	0.014	0.325	0.007	308.54	0.47
AS19	0.68	0.004	0.67	0.008	678.95	6.61	1.96E62	6.20E62	2.64E-4	0.013	0.324	0.005	308.58	0.42
AS20	0.68	0.003	0.67	0.005	677.95	5.49	3.12E44	6.56E44	0.003	0.008	0.329	0.016	308.61	0.54
AS21	0.67	0.005	0.66	0.007	709.25	7.59	1.55E50	4.91E50	−0.003	0.012	0.323	0.003	315.18	0.39

(continued)

Table 4.8 (continued)

PS	$H(\varepsilon)$		$M(\varepsilon)$		L_{opt}		ε^*		r_{FD}		τ		δ	
	Mean	Std.	Mean	Std.	Mean	Std.	Mean	Std.	Mean	Std.	Mean	Std.	Mean	Std.
AS22	0.66	0.005	0.67	0.006	793.91	5.47	9.28E50	2.93E51	−6.70E-4	0.013	0.322	0.009	331.38	0.35
AS23	0.67	0.005	0.67	0.003	843.65	6.31	2.12E45	6.61E45	−0.010	0.013	0.317	0.004	340.64	0.42
AS24	0.67	0.005	0.66	0.005	877.56	8.50	1.03E56	3.28E56	0.002	0.012	0.319	0.010	346.73	0.46
AS25	0.66	0.003	0.66	0.004	944.42	5.86	8.14E53	2.57E54	−0.003	0.010	0.313	0.002	358.66	0.53
AS26	0.66	0.004	0.66	0.005	947.70	9.03	4.06E44	9.80E44	−0.003	0.012	0.318	0.008	358.87	0.44
AS27	0.66	0.005	0.67	0.003	946.13	4.17	5.34E54	1.68E55	−0.002	0.009	0.312	0.001	358.69	0.33
AS28	0.66	0.003	0.66	0.006	947.06	9.37	5.10E46	1.60E47	0.001	0.009	0.317	0.009	358.81	0.41
AS29	0.66	0.003	0.66	0.004	949.18	6.87	4.66E66	1.47E67	−1.08E-4	0.015	0.318	0.011	358.77	0.34
AS30	0.66	0.003	0.67	0.005	947.42	6.61	6.18E47	1.95E48	0.001	0.010	0.313	0.005	358.58	0.25
AS31	0.66	0.004	0.66	0.004	947.15	6.06	8.97E47	1.77E48	0.001	0.010	0.318	9.39E-5	358.68	0.38
AS32	0.66	0.003	0.67	0.004	946.76	5.74	4.14E56	1.31E57	0.005	0.013	0.315	0.008	358.84	0.38
AS33	0.66	0.004	0.66	0.003	948.53	7.46	4.85E55	1.53E56	−0.006	0.014	0.314	0.006	358.77	0.22
AS34	0.66	0.003	0.66	0.003	947.68	9.17	3.41E54	1.08E55	−0.001	0.008	0.320	0.011	358.92	0.35
AS35	0.66	0.003	0.66	0.006	946.63	6.67	3.68E59	1.06E60	−0.003	0.011	0.314	0.005	358.73	0.36

Table 4.9 Results of information and statistical measures for 30 real protein landscapes in 3D AB off-lattice model

PS	$H(\varepsilon)$		$M(\varepsilon)$		L_{opt}		ε^*		r_{FD}		τ		δ	
	Mean	Std.	Mean	Std.	Mean	Std.	Mean	Std.	Mean	Std.	Mean	Std.	Mean	Std.
RS1	0.67	0.007	0.65	0.012	140.06	2.60	9.72E47	3.05E48	0.004	0.026	0.385	0.009	147.21	0.32
RS2	0.67	0.009	0.66	0.009	140.86	2.00	1.19E42	3.48E42	−0.005	0.029	0.386	0.009	147.26	0.35
RS3	0.65	0.007	0.66	0.012	183.20	3.84	2.33E44	7.39E44	−0.011	0.023	0.372	0.003	167.73	0.47
RS4	0.66	0.007	0.66	0.013	196.67	3.30	1.60E47	5.06E47	0.003	0.025	0.368	0.004	173.54	0.48
RS5	0.67	0.005	0.66	0.009	210.38	2.99	2.14E41	6.47E41	0.005	0.020	0.365	0.002	179.56	0.37
RS6	0.68	0.008	0.66	0.008	254.78	3.51	7.01E49	2.21E50	−0.002	0.022	0.363	0.011	196.62	0.39
RS7	0.68	0.010	0.66	0.007	256.65	3.21	3.31E54	1.04E55	0.017	0.024	0.358	0.002	196.49	0.34
RS8	0.68	0.007	0.66	0.007	298.33	3.24	9.16E52	2.89E53	0.006	0.024	0.355	0.010	211.92	0.53
RS9	0.67	0.007	0.66	0.009	317.98	5.888	2.82E61	8.91E61	0.003	0.014	0.359	0.013	216.64	0.34
RS10	0.67	0.007	0.66	0.007	314.48	2.92	1.67E43	5.23E43	−0.004	0.024	0.354	0.010	216.71	0.30
RS11	0.67	0.009	0.66	0.009	319.15	5.77	8.69E47	2.74E48	0.001	0.020	0.360	0.013	216.59	0.45
RS12	0.67	0.006	0.66	0.008	328.10	2.85	3.39E46	1.07E47	0.011	0.020	0.352	0.012	221.54	0.77
RS13	0.66	0.003	0.66	0.008	376.29	3.91	1.05E51	3.33E51	−0.004	0.016	0.346	0.009	235.22	0.43
RS14	0.66	0.004	0.66	0.008	376.67	5.20	8.24E43	2.60E44	−0.007	0.020	0.347	0.010	235.28	0.27
RS15	0.65	0.006	0.67	0.014	374.73	6.63	4.36E47	1.15E48	−0.007	0.018	0.351	0.014	235.37	0.36
RS16	0.66	0.006	0.66	0.009	374.01	5.82	2.32E55	7.34E55	−0.013	0.023	0.352	0.016	235.36	0.32
RS17	0.65	0.004	0.66	0.009	375.66	3.46	1.01E52	3.21E52	2.39E-4	0.020	0.349	0.013	235.28	0.32
RS18	0.66	0.005	0.66	0.010	452.27	7.24	8.81E44	2.78E45	0.012	0.016	0.339	0.006	256.48	0.50
RS19	0.67	0.005	0.66	0.006	501.12	5.05	1.95E45	4.55E45	7.73E-4	0.009	0.336	0.008	268.77	0.46
RS20	0.67	0.005	0.66	0.005	499.47	3.54	2.47E45	6.96E45	−0.005	0.014	0.340	0.010	268.62	0.42

(continued)

Table 4.9 (continued)

| PS | $H(\varepsilon)$ | | $M(\varepsilon)$ | | L_{opt} | | ε^* | | r_{FD} | | τ | | δ | |
	Mean	Std.	Mean	Std.	Mean	Std.	Mean	Std.	Mean	Std.	Mean	Std.	Mean	Std.
RS21	0.67	0.005	0.66	0.007	500.88	4.77	1.86E44	5.88E44	0.003	0.010	0.335	0.005	268.73	0.53
RS22	0.67	0.008	0.66	0.006	515.98	7.38	8.88E47	2.80E48	−0.002	0.019	0.334	0.006	272.66	0.49
RS23	0.67	0.004	0.66	0.008	645.08	7.87	1.14E54	3.20E54	0.006	0.012	0.324	2.80E-4	301.70	0.49
RS24	0.66	0.006	0.66	0.003	780.23	6.14	7.35E46	2.32E47	−9.25E-4	0.010	0.322	0.006	328.29	0.37
RS25	0.65	0.004	0.66	0.005	874.33	6.33	2.27E59	6.95E59	−0.006	0.011	0.317	0.006	346.83	0.37
RS26	0.66	0.006	0.66	0.006	949.11	8.56	6.24E50	1.97E51	0.004	0.013	0.316	0.070	358.67	0.47
RS27	0.67	0.002	0.66	0.005	1139.06	7.37	2.14E47	3.43E47	0.001	0.010	0.307	0.002	389.47	0.26
RS28	0.67	0.003	0.66	0.003	1291.08	7.97	2.20E55	6.96E55	2.56E-4	0.012	0.305	0.007	412.90	0.31
RS29	0.68	0.003	0.66	0.003	1349.83	8.04	4.84E63	1.53E64	0.004	0.010	0.306	0.012	420.43	0.36
RS30	0.68	0.002	0.66	0.004	1542.61	10.42	2.02E66	6.40E66	−0.003	0.008	0.297	2.97E-5	447.07	0.36

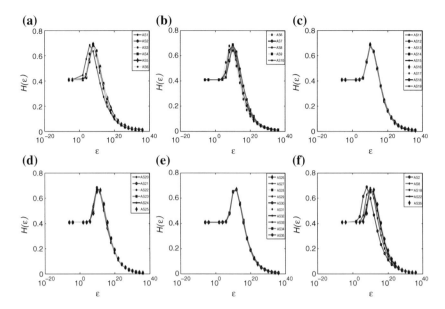

Fig. 4.14 Mean entropy measure, $H(\varepsilon)$ curves for artificial protein sequences in 3D AB off-lattice model: **a** AS1 - AS6 $\in G_{A_1}$, **b** AS6 - AS10 $\in G_{A_2}$, **c** AS11 - AS19 $\in G_{A_3}$, **d** AS20 - AS25 $\in G_{A_4}$, **e** AS26 - AS35 $\in G_{A_5}$ and **f** AS2, AS8, AS18, AS22 and AS35 taken from each group

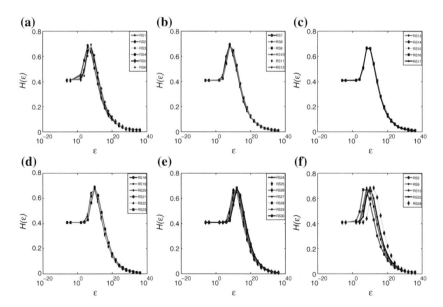

Fig. 4.15 Mean entropy measure, $H(\varepsilon)$ curves for real protein sequences in 3D AB off-lattice model: **a** RS1 - RS6 $\in G_{R_1}$, **b** RS7 - RS12 $\in G_{R_2}$, **c** RS13 - RS17 $\in G_{R_3}$, **d** RS18 - RS23 $\in G_{R_4}$, **e** RS24 - RS30 $\in G_{R_5}$ and **a** RS3, RS9, RS15, RS22 and RS28 taken from each group

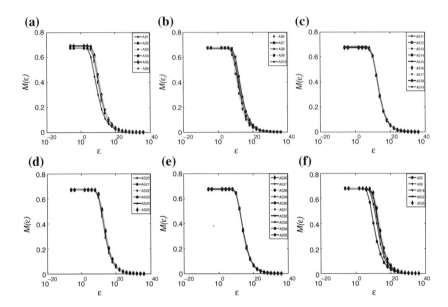

Fig. 4.16 Mean partial information content measure, $M(\varepsilon)$ curves for artificial protein sequences in 3D AB off-lattice model: **a** AS1 - AS6 $\in G_{A_1}$, **b** AS6 - AS10 $\in G_{A_2}$, **c** AS11 - AS19 $\in G_{A_3}$, **d** AS20 - AS25 $\in G_{A_4}$, **e** AS26 - AS35 $\in G_{A_5}$ and **f** AS2, AS8, AS18, AS22 and AS35 taken from each group

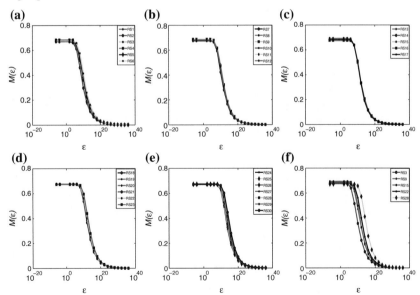

Fig. 4.17 Mean partial information content measure, $M(\varepsilon)$ curves for real protein sequences in 3D AB off-lattice model: **a** RS1 - RS6 $\in G_{R_1}$, **b** RS7 - RS12 $\in G_{R_2}$, **c** RS13 - RS17 $\in G_{R_3}$, **d** RS18 - RS23 $\in G_{R_4}$, **e** RS24 - RS30 $\in G_{R_5}$ and **a** RS3, RS9, RS15, RS22 and RS28 taken from each group

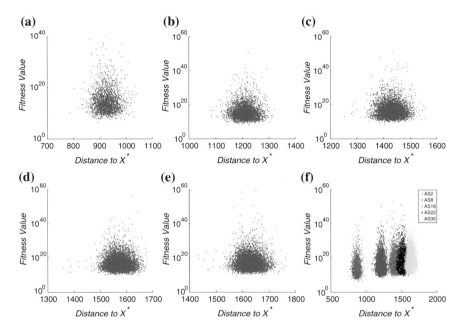

Fig. 4.18 Scatter plots of fitness versus distance to the global minimum (X^*) for artificial protein sequences in 3D AB off-lattice model: **a** AS4 with lenght 23, **b** AS7 with lenght 34, **c** AS21 with lenght 50, **d** AS24 with lenght 60, **e** AS34 with lenght 64 and **f** AS2, AS8, AS18, AS22 and AS35 taken from each group

as the sequence lengths increase. For all the real protein sequences, $H(\varepsilon)$ values are approximately close to 0.7 which demonstrates high rugged landscape.

Figure 4.14 reveals that all the artificial protein sequences are attained same initial $H(\varepsilon)$ value close to 0.4 at $\varepsilon = 10^{-6}$. After that $H(\varepsilon)$ value increases up to certain value of ε and then decreasing as ε increases. However, sequences attained maximum $H(\varepsilon)$ at different values of ε which shows increasing trend as sequence length increases. $H(\varepsilon)$ finally converges to 0 at $\varepsilon = 10^{36}$, indicates flat landscape structure. We also observe similar patterns for 3D real protein sequences, shown in Fig. 4.15.

The partial information content $M(\varepsilon)$ of the landscape estimates the modality of the landscape path. Tables 4.8 and 4.9 presents the measures of $M(\varepsilon)$ for all protein sequences, roughly constant and close to 0.7. This indicates existence of high modality in the landscape structure. Therefore, high modality is encountered on the protein landscape path irrespective of the protein sequence lengths in 3D AB off-lattice model. Figures 4.16 and 4.17 presents characteristics of $M(\varepsilon)$ measure over the different values of ε.

It is surprising to observe in Tables 4.8 and 4.9 that L_{opt} significantly increases as lengths of the protein sequence increases, independent of its type. Therefore, most important structural features like the existence of modality and local optima of a protein landscape are extracted using L_{opt}. Measures of information stability, ε^* are

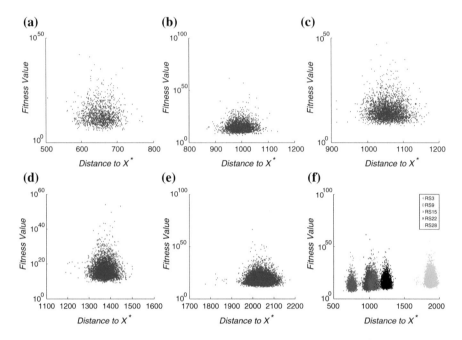

Fig. 4.19 Scatter plots of fitness versus distance to the global minimum (X^*) for real protein sequence in 3D AB off-lattice model: **a** RS1 with lenght 13, **b** RS10 with lenght 25, **c** RS15 with lenght 29, **d** RS23 with lenght 46, **e** RS30 with lenght 98 and **f** RS3, RS9, RS15, RS22 and RS28 taken from each group

shown in Tables 4.8 and 4.9 for all the protein instances and does not follow any particular pattern as sequence lengths increase. However, ε^* attained a large value in both the artificial and real protein instances. Basically, ε^* is highly dependent on sampling technique and fitness function rather than landscape structure which is evident from high ε^* values.

From Table 4.8, it is clear that we obtained typical small positive values of r_{FD} on 16 ASs with an exception on 19 ASs which attained small negative value. On the other hand, r_{FD} attained small positive value on 17 RSs while small negative value attained on 13 RSs as observed in Table 4.9. However, all other r_{FD} values are around 0 and not follow any pattern as sequence length increases either in artificial or real protein. This implies that there is no correlation between the fitness of the candidate solutions and corresponding distances to the global minimum. Therefore, r_{FD} measure provides complex landscape structures which are needle-in-a-haystack without any global topology regardless of the type of protein sequence. Figures 4.18 and 4.19 present the fitness distance scatter plots for artificial and real protein sequences based on 3D AB off-lattice model. In all sub-figures of Fig. 4.18, the sample points are scattered in outer regions suggesting a complete absence of correlation. There is no difference in overall shape and trends of samples for both the artificial and real protein sequences, which is obvious since the points in respective scatter plots (Figs. 4.18 and 4.19) are highly uncorrelated.

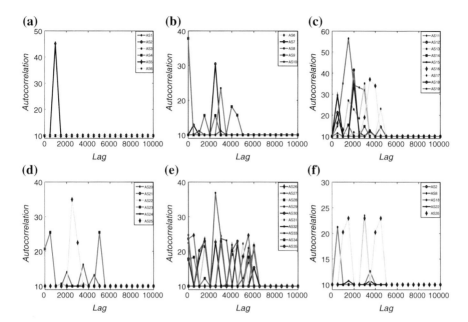

Fig. 4.20 Correlograms for artificial protein sequences in 3D AB off-lattice model: **a** AS1 - AS6 $\in G_{A_1}$, **b** AS6 - AS10 $\in G_{A_2}$, **c** AS11 - AS19 $\in G_{A_3}$, **d** AS20 - AS25 $\in G_{A_4}$, **e** AS26 - AS35 $\in G_{A_5}$ and **f** AS2, AS8, AS18, AS22 and AS35 taken from each group

The correlograms for all the protein instances are depicted in Figs. 4.20 and 4.21. In Fig. 4.20a, autocorrelation values are around 10 for ASs in G_{A_1} over different lag with an exception on AS5 at 1500 lag value. Remaining sequences are attained reasonable autocorrelation values around 10 and sometimes fluctuate at different lag value. Smaller value acknowledge uncorrelated neighboring points w.r.t. fitness in the landscape which reveals dissimilarity among the points. Dissimilar sample points in a landscape form a rugged landscape structure. Dissimilarity increases with increase of protein sequence lengths. Figure 4.21 presents correlograms for real protein sequences showing autocorrelation values close to 10 over different lag as sequence length increases with few fluctuations at different lag value. An unexpected pattern is observed with lag values 4000, 4500, 5000 and 5500 for RS28. On the other hand, correlation length (τ) defines a distance by which two points become uncorrelated. Results are given in Table 4.8 where τ decreases with increase of sequence length for artificial sequences. This observation is also noticed on real protein sequences as shown in Table 4.9, with very small (τ) which implies uncorrelated neighboring sample points.

From Tables 4.8 and 4.9, it is clear that dispersion (δ) values are monotonically increasing for both the artificial and real protein sequences as length increases. It has been observed that measure of δ increases with sequence lengths regardless the type of the protein sequence. Moreover, significant change exists from sequence

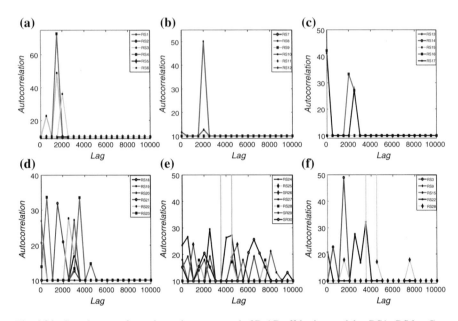

Fig. 4.21 Correlograms for real protein sequences in 3D AB off-lattice model: **a** RS1 - RS6 $\in G_{R_1}$, **b** RS7 - RS12 $\in G_{R_2}$, **c** RS13 - RS17 $\in G_{R_3}$, **d** RS18 - RS23 $\in G_{R_4}$, **e** RS24 - RS30 $\in G_{R_5}$ and **a** RS3, RS9, RS15, RS22 and RS28 taken from each group

to sequence. Large value of dispersion (δ) indicates presence of multi-funnels in the protein landscape and best solutions are far away from possible solutions. It is possible only when a significant change is made on the basin shape that consists clustered local optima in a landscape. Therefore, dispersion metric captures an important structural feature i.e. presence of multi-funnels from the protein landscape that highly influence the performance of any optimization algorithm. Figures 4.22 and 4.23 present dispersion characteristics with increase of sample size for the artificial and real protein sequences.

Tables 4.10 and 4.11 show Pearson's correlation coefficient ($\rho_{x,y}$) for the artificial and real protein instances in 3D AB off-lattice model. A strong correlation ($\rho_{x,y} \geq 0.7$) between the measures is shown by boldface. In Table 4.10, focusing on the results of $H(\varepsilon)$, it is surprising to note that no measure is correlated with $H(\varepsilon)$ and provides same results for all the protein instances. We observed that the measures L_{opt} and δ are strongly correlated ($\rho_{x,y} \approx 1$) to each other, therefore, any one of them is sufficient for analyzing the landscape. However, correlation (Table 4.10) indicates that most of the measures ($H(\varepsilon)$, $M(\varepsilon)$, ε^*, r_{FD}, τ) are not correlated to each other. In case of the real protein sequences, $H(\varepsilon)$ is strongly correlated with $M(\varepsilon)$, τ and δ, shown in Table 4.11. Moreover, from Table 4.11, it is clear that most of the measures $M(\varepsilon)$, L_{opt}, ε^*, r_{FD} and τ are not correlated to each other. Therefore, extracting structural features of the protein landscape in 3D AB off-lattice model requires maximum number of landscape measures.

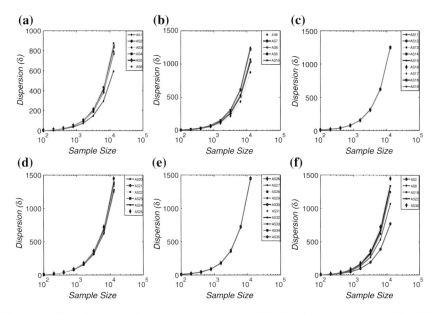

Fig. 4.22 Dispersion for artificial protein sequences in 3D AB off-lattice model: **a** AS1 - AS6 $\in G_{A_1}$, **b** AS6 - AS10 $\in G_{A_2}$, **c** AS11 - AS19 $\in G_{A_3}$, **d** AS20 - AS25 $\in G_{A_4}$, **e** AS26 - AS35 $\in G_{A_5}$ and **f** AS2, AS8, AS18, AS22 and AS35 taken from each group

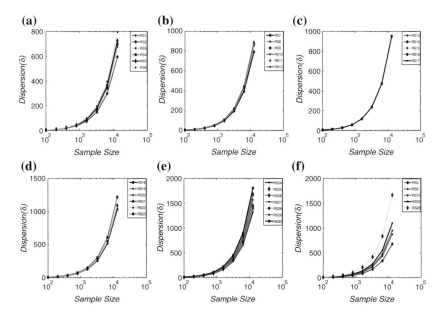

Fig. 4.23 Dispersion for real protein sequences in 3D AB off-lattice model: **a** RS1 - RS6 $\in G_{R_1}$, **b** RS7 - RS12 $\in G_{R_2}$, **c** RS13 - RS17 $\in G_{R_3}$, **d** RS18 - RS23 $\in G_{R_4}$, **e** RS24 - RS30 $\in G_{R_5}$ and **a** RS3, RS9, RS15, RS22 and RS28 taken from each group

Table 4.10 Pearson correlation coefficients ($\rho_{x,y}$) between landscape measures for all artificial protein sequences in 3D AB off-lattice model

Measure	$M(\varepsilon)$	L_{opt}	ε^*	r_{FD}	τ	δ
$H(\varepsilon)$	−0.357	−0.646	−0.136	0.336	0.484	−0.576
$M(\varepsilon)$		−0.173	−0.999	0.009	0.106	−0.152
L_{opt}			0.182	−0.328	−0.948	**0.990**
ε^*				−0.012	−0.114	0.161
r_{FD}					0.322	−0.323
τ						−0.977

Table 4.11 Pearson correlation coefficients ($\rho_{x,y}$) between landscape measures for all real protein sequences in 3D AB off-lattice model

Measure	$M(\varepsilon)$	L_{opt}	ε^*	r_{FD}	τ	δ
$H(\varepsilon)$	**0.999**	0.480	0.076	0.037	**0.983**	**0.778**
$M(\varepsilon)$		0.517	0.118	0.051	−0.592	0.576
L_{opt}			0.506	−0.020	−0.936	0.985
ε^*				−0.101	−0.373	0.433
r_{FD}					−0.022	−0.019
τ						−0.980

4.5 Performance Analysis of the Metaheuristic Algorithms

In this section, the performance of six real-coded optimization algorithms is analyzed in 2D and 3D AB off-lattice model using different lengths of ASs and RSs with the same system specifications and protein instances.

4.5.1 Simulation Configurations

In this study, six optimization algorithms with different characteristics have been considered in order to select the most appropriate algorithm with a feedback from the performance analysis based on the structural features of the PSP problem. The algorithms that we consider are real coded GA [167], DE [47, 48], PSO [54], ABC [168], CMA-ES [169] and Bees Algorithms [59]. The algorithms are chosen due to great relevance in evolutionary computation and continuous growth in diverse fields of applications. We have considered the standard versions of each algorithm in terms of its simplicity and popularity. However, different variants of these algorithms provide better results than the standard version. It is very important to mention that the aim is to explain the algorithmic performance based on the structural features and obtain guidelines for most appropriate algorithms and not to produced any kind of variants of an algorithm for the PSP problem. The representation scheme for all the algorithms is same in order to conform uniform conditions. An ith candidate solution is encoded as a D-dimensional vector of decision variables $\mathbf{x}_i = (x_{i1}, x_{i2}, \ldots, x_{iD}) \in \mathbb{R}^D$ over the continuous search space $U = [-180°, 180°]^D$. For each algorithm, the initial population is generated using uniformly distributed random numbers across the search space given as follows.

$$x_{ij} = x_j^{min} + (x_j^{max} - x_j^{min})rand(0, 1);$$
$$i = 1, 2, \ldots, N \quad and \quad j = 1, 2, \ldots, D,$$

(4.11)

where x_j^{min} and x_j^{max} denote the lower and upper bound of the jth component of the ith candidate solution $\mathbf{x}_i$. $rand(0, 1)$ is a uniformly distributed random number between 0 and 1. The stopping criterion depends on the maximum number of function evaluations (FEs) defined as $Max\_FEs = (10,000 \times L)/N$ (L is the length of protein sequence). Each algorithms are executed 30 independent runs to measure robustness.

Since the landscape features of the energy functions used for 2D and 3D AB off-lattice models indicate multi-modality, ruggedness and deceptiveness (please see Tables 4.17 and 4.18 for a summary of these features in Sect. 4.4), to make the comparison fair enough we have set up the parameters accordingly. For example, GA uses a Roulette wheel selection operator, random single point crossover operator and uniform mutation operator with crossover ratio (p_c) 0.5, mutation probability (P_m) 0.1 and population size is 50. These are the standard GA configurations used earlier in literature for optimizing multi-modal benchmark functions like the Shekel's foxholes problem. The high mutation probability of 0.1 makes random perturbations in the chromosomes thereby helping them to jump out from the local optima occasionally. In DE algorithm, the less greedy offspring generation scheme DE/rand/1/bin is used to reduce attraction to specific points in the search space. This helps in better exploration and can avoid trapping at a local optimal basin. To facilitate exploitation, the crossover probability (CR) is set to 0.3 so that on the average more components may be inherited from the parent. The scale factor (F) is randomly generated within $[0.5, 1]$ for each candidate solution in each generation. Such randomization, called dither, helps the solutions to quickly explore a multi-modal and rugged search space as indicated in [170]. The PSO algorithm is executed with the inertia weight ω, decreasing linearly as the generation count increases in the following way:

$$\omega = 0.9 - (0.9 - 0.4)\frac{t}{t_{max}},$$

(4.12)

where t and t_{max} represent the current generation and the maximum number of generations are allowed, respectively. Such decreasing inertia weight facilitates greater diversity during the initial phases of search and intensification of the search during the later stages. This feature helps PSO to perform better on multi-modal landscapes. Similarly, the control parameters for ABC, BA, and CMA-ES are chosen from the existing literature (indicated in Table 4.12) so as to achieve standard performance on

Table 4.12 Parametric set-up for the algorithms compared

GA [167]		DE [48]		PSO [54]	
Parameter	Value	Parameter	Value	Parameter	Value
Population size	50	Population size	100	Population size	50
Crossover prob. (p_c)	0.5	Mutation strategy	DE/rand/1/bin	Acceleration coefficients (c_1 and c_2)	1.496
Mutation prob. (P_m)	0.1	Crossover prob. (CR)	0.3	Inertia weight. (ω)	$\omega = 0.9 - (0.9 - 0.4) \times \frac{t}{t_{max}}$
		Scale factor (F)	Randomly generated within [0.5, 1]		
ABC [168]		CMA-ES [169]		BA [59]	
Parameter	Value	Parameter	Value	Parameter	Value
Population size	50	Population size λ	50	Population size	50
Scout bee limit	D	Recombination parents ($\frac{\lambda}{2}$)	25	Elite solutions (ne)	10
		Standard deviation	$\sigma = 0.3$	Best solutions (nb)	30
				Recruited bees for elite solutions (nre)	10
				Recruited bees for remaining best solutions (nrb)	5
				Neighborhood structure	As in [171]

multi-modal, rugged and deceptive landscapes as much as possible. The parametric settings of the comparison algorithms are summarized in Table 4.12.

4.5.2 Performance Measures

We report the results of *best*, *mean* and *standard deviation (std.)* of the best-of-the run over 30 independent runs for each algorithm on ASs and RSs using 2D and 3D AB off-lattice model in Tables 4.13, 4.14, 4.15 and 4.16. Wilcoxon's rank sum test for independent runs [172] is conducted in order to judge whether the *mean* results obtained with the best performing algorithm differ from the *mean* results of the rest of the competitors in a statistically significant way. "+" symbol in Tables 4.13, 4.14, 4.15 and 4.16 indicate that the best algorithm in row yields significantly better results compared to the corresponding algorithm while "$\approx$" indicate that there is no significance difference between the best algorithm and the corresponding compared algorithm in the statistical sense. Minimum on the *mean* results for each algorithm for the protein instances are in bold face while minimum on the *best* results are underlined shown in Tables 4.13, 4.14, 4.15 and 4.16.

4.5.3 Results and Discussion

4.5.3.1 For 2D AB Off-Lattice Model

The performance metric results of the optimization algorithms are shown in Tables 4.13 and 4.14 while landscape measure results are given in Tables 4.4 and 4.5 for 2D AB off-lattice model with artificial and real protein sequences. The landscape measure results are analyzed to infer that the features are highly sensitive to the performance of the GA algorithm. GA achieves best on the *best* results for smaller lengths of ASs (AS1 and AS5) but surprisingly it is not true for RSs. However, GA is almost equal to DE on *mean* results for relatively shorter lengths of RSs compared to the other approaches. GA has decreased performance in *best* or *mean* results for higher lengths of protein sequences irrespective of ASs or RSs. GA has high variability in each runs on some of the protein sequences excepting BA as indicated by the *standard deviation* results. As evident from the Tables 4.4 and 4.5, the local optima, lack of correlation between sample points and existence of funnels are increase as the length of protein sequence increases. Therefore, GA has poor performance on the landscape structure of the higher lengths protein instances in 2D AB off-lattice model.

DE offers good *mean* results for AS1, RS1-RS4 and RS6 and almost same results are obtained by the GA and PSO algorithms. No significant improved performances are observed in ASs excepting AS1 by the DE algorithm irrespective of the protein sequence lengths. On AS3, AS5, AS6, RS5, RS7, and RS9 the *mean* results of DE are quite closer to the best results obtained by the PSO algorithm. Overall, DE performances are decreased as protein sequence length increases. Although, it has less variability during the runs for all the ASs and 27 out of the 30 RSs. Thus, the selected configuration of the DE algorithm fails to solve the PSP problem which consists many local optima, multi-funnels and highly rugged in their landscape structures. In general, a success of DE highly depends on mutation and crossover operators which can be controlled by the parameters scale factor (F) and crossover probability (CR). For solving PSP problem successfully by DE, we need to tune the parameters such as F and CR as well as modifying or incorporating new mutation and crossover operator in order to prevent premature convergence.

The PSO algorithm offered superior performance in both *best* and *mean* results on the majority of ASs (AS3, AS4, AS6-AS12, AS14-AS21, AS24, AS26, AS28, AS31-AS33) and RSs (RS5, RS7-RS28) compared to the other algorithms. PSO has comparable performance on the sequences AS1, AS2, AS5, AS13, AS29 and AS30

Table 4.13 Results for six optimization algorithms on 35 artificial protein sequences using 2D AB off-lattice model

PS No.	PSL	GA			DE			PSO		
		Best	Mean	Std.	Best	Mean	Std.	Best	Mean	Std.
AS1	13	−2.248	−1.721 (≈)	0.467	−2.275	**−2.274**	0.001	−1.806	−1.672 (≈)	0.166
AS2	20	−7.743	**−6.551**	1.362	−5.844	−5.797 (≈)	0.049	−6.669	−6.420 (≈)	0.217
AS3	21	−3.188	−3.020 (+)	0.164	−3.856	−3.667 (≈)	0.169	−4.467	**−4.273**	0.273
AS4	23	−6.374	−5.917 (≈)	0.484	−5.532	−5.295 (+)	0.212	−8.174	**−6.584**	1.381
AS5	24	−5.014	−3.913 (≈)	0.980	−3.275	−3.153 (≈)	0.131	−4.417	−3.340 (≈)	1.118
AS6	25	−5.441	−3.092 (+)	2.216	−5.113	−4.937 (≈)	0.167	−7.364	**−5.023**	2.441
AS7	34	−5.078	−4.023 (+)	1.386	−4.412	−4.132 (+)	0.292	−7.220	**−6.044**	1.158
AS8	36	−9.559	−8.766 (+)	0.713	−8.430	−7.908 (+)	0.660	−13.536	**−10.350**	2.787
AS9	46	−19.548	−17.179 (+)	3.634	−8.684	−8.542 (+)	0.162	−22.986	**−20.487**	2.783
AS10	48	−13.31	−11.118 (+)	1.937	−10.069	−8.950 (+)	0.983	−20.429	**−18.172**	2.257
AS11	48	−12.698	−11.179 (+)	1.505	−7.480	−7.281 (+)	0.250	−20.396	**−18.379**	1.836
AS12	48	−14.239	−11.871 (+)	2.194	−9.142	−8.985 (+)	0.224	−19.336	**−17.603**	2.263
AS13	48	−15.224	−11.680 (+)	3.070	−8.391	−8.172 (+)	0.338	−15.313	−14.906 (≈)	0.623
AS14	48	−17.409	−13.265 (+)	5.216	−7.102	−6.513 (+)	0.607	−17.770	**−17.452**	0.323
AS15	48	−17.051	−15.402 (+)	1.902	−9.525	−9.032 (+)	0.651	−22.677	**−20.444**	2.008
AS16	48	−18.697	−15.758 (+)	2.737	−9.583	−9.314 (+)	0.246	−19.806	**−18.491**	2.261
AS17	48	−15.907	−14.435 (+)	2.003	−10.069	−8.633 (+)	1.283	−18.624	−15.880 (≈)	2.378
AS18	48	−12.882	−12.172 (+)	0.629	−8.723	−7.996 (+)	0.632	−19.821	**−18.538**	3.128
AS19	48	−14.468	−12.069 (+)	2.209	−8.012	−7.739 (+)	0.237	−20.067	**−18.574**	2.321
AS20	48	−9.912	−9.087 (+)	1.045	−7.649	−6.878 (+)	0.938	−15.055	−14.076 (≈)	0.921

(continued)

Table 4.13 (continued)

PS No.	PSL	GA			DE			PSO		
		Best	Mean	Std.	Best	Mean	Std.	Best	Mean	Std.
AS21	50	−15.368	−13.987 (+)	1.343	−7.870	−7.244 (+)	0.596	−23.499	**−19.110**	4.953
AS22	55	−7.160	−7.026 (≈)	0.210	−4.949	−4.353 (+)	0.592	−7.087	−5.645 (+)	2.132
AS23	58	−15.199	−13.154 (+)	1.781	−5.964	−5.700 (+)	0.269	−15.504	−14.432 (+)	1.072
AS24	60	−38.839	−38.108 (+)	0.700	−11.932	−11.659 (+)	0.420	−45.894	**−41.122**	4.314
AS25	64	−23.743	−18.902 (+)	4.310	−9.360	−7.569 (+)	1.943	−29.194	−28.527 (+)	0.587
AS26	64	−8.458	−7.720 (+)	0.809	−2.351	−2.127 (+)	0.223	−14.072	**−13.018**	1.054
AS27	64	−11.809	−8.615 (+)	2.835	−4.7734	−4.326 (+)	0.478	−14.147	−12.134 (+)	2.012
AS28	64	−23.421	−20.400 (+)	4.982	−6.967	−5.420 (+)	1.525	−26.787	**−24.864**	2.218
AS29	64	−10.702	−6.703 (+)	3.533	−3.568	−3.016 (+)	0.576	−12.802	−11.956 (≈)	1.3687
AS30	64	−13.888	−10.946 (+)	3.963	−3.568	−3.059 (+)	0.651	−17.149	−15.902 (≈)	1.095
AS31	64	−13.167	−10.129 (+)	2.850	−4.186	−3.477 (+)	0.618	−17.653	**−15.060**	2.605
AS32	64	−13.158	−9.199 (+)	4.581	−3.793	−3.006 (+)	0.913	−15.808	**−15.021**	0.787
AS33	64	−13.200	−10.810 (+)	2.351	−4.462	−3.518 (+)	1.007	−14.766	**−13.150**	0.616
AS34	64	−14.630	−12.048 (+)	3.127	−4.733	−4.287 (+)	0.443	−14.0816	−13.195 (+)	0.953
AS35	64	−6.920	−5.678 (+)	1.329	−2.109	−1.134 (+)	1.617	−8.649	−6.180 (+)	2.351

Table 4.13 (continued)

PS No.	PSL	ABC			CMA-ES			BA		
		Best	Mean	Std.	Best	Mean	Std.	Best	Mean	Std.
AS1	13	−2.153	−2.020 (≈)	0.118	−1.483	−1.303 (+)	0.156	−0.865	−0.110 (+)	0.665
AS2	20	−6.136	−5.894 (≈)	0.216	−2.769	−2.456 (+)	0.328	−0.842	0.561 (+)	1.644
AS3	21	−3.134	−2.943 (+)	0.189	−2.747	−2.455 (+)	0.262	0.149	1.777 (+)	1.983
AS4	23	−5.658	−5.396 (+)	0.227	−2.704	−2.678 (+)	0.025	1.332	13.363 (+)	19.548
AS5	24	−4.683	**−3.933**	0.713	−0.866	−0.603 (+)	0.232	1.185	5.018 (+)	5.080
AS6	25	−5.122	−4.884 (≈)	0.251	−0.450	−0.423 (+)	0.037	1.362	10.600 (+)	9.839
AS7	34	−5.966	−4.865 (+)	0.965	−3.080	−2.793 (+)	0.300	55.060	99.00 E02 (+)	14.74 E03
AS8	36	−9.789	−9.119 (+)	0.701	−4.442	−4.263 (+)	0.250	64.979	13.99E02 (+)	22.88E02
AS9	46	−19.782	−19.188 (+)	0.520	−5.543	−4.761 (+)	0.684	16.7E03	29.31E04 (+)	39.51E04
AS10	48	−16.939	−16.048 (+)	0.786	−7.200	−6.587 (+)	0.628	43.67 E03	25.90 E04 (+)	27.89 E04
AS11	48	−16.016	−15.187 (+)	0.791	−3.072	−2.661 (+)	0.446	82.05 E03	15.54 E04 (+)	78.02 E03
AS12	48	−15.506	−15.248 (+)	0.248	−3.441	−3.372 (+)	0.072	69.28 E03	27.60 E04 (+)	18.10 E04
AS13	48	−15.657	**−14.944**	1.222	−4.915	−4.397 (+)	0.450	12.59E03	27.60 E04 (+)	30.10 E04
AS14	48	−16.424	−15.443 (+)	0.859	−5.712	−4.326 (+)	1.211	98.78E03	23.03 E04 (+)	15.90 E04
AS15	48	−17.931	−17.290 (+)	0.603	−4.785	−4.591 (+)	0.251	35.17E034	96.29 E03 (+)	53.06 E03
AS16	48	−17.677	−17.209 (+)	0.559	−4.657	−4.558 (+)	0.097	77.41E02	26.72 E04 (+)	38.30 E04
AS17	48	−17.019	**−16.057**	0.881	−4.595	−3.543 (+)	0.912	11.40E01	22.24 E04 (+)	28.65 E04
AS18	48	−18.982	−17.140 (+)	1.409	−5.533	−4.269 (+)	1.099	26.33E03	48.77 E04 (+)	43.72 E04
AS19	48	−16.134	−15.698 (+)	0.542	−6.057	−5.073 (+)	0.866	10.92E04	39.79 E04 (+)	48.36 E04

(continued)

Table 4.13 (continued)

PS No.	PSL	ABC			CMA-ES			BA		
		Best	Mean	Std.	Best	Mean	Std.	Best	Mean	Std.
AS20	48	−14.881	**−14.351**	0.532	−4.784	−4.383 (+)	0.352	15.46E04	28.53 E04 (+)	18.99E04
AS21	50	−19.033	−17.762 (+)	1.123	−7.789	−6.301 (+)	1.290	29.58E03	18.80E04 (+)	18.34E04
AS22	55	−8.260	**−7.796**	0.440	−4.031	−3.539 (+)	0.434	10.40E04	13.61 E05 (+)	11.58E05
AS23	58	−18.215	**−17.645**	0.576	−4.215	−3.218 (+)	0.874	13.45E04	26.59 E05 (+)	22.01E05
AS24	60	−39.354	−38.429 (+)	1.095	−13.322	−12.294 (+)	0.937	50.68E04	41.61 E05 (+)	60.57E05
AS25	64	−31.195	**−30.148**	1.515	−10.918	−10.308 (+)	0.831	14.39E04	32.51 E05 (+)	31.78E05
AS26	64	−11.341	−11.177 (+)	0.164	−2.092	−1.862 (+)	0.199	37.31E04	58.50 E05 (+)	58.84E05
AS27	64	−14.820	**−13.273**	1.408	−4.349	−3.502 (+)	0.738	14.60E04	68.07E05 (+)	74.52E05
AS28	64	−22.634	−21.893 (+)	0.905	−6.062	−5.533 (+)	0.489	43.60E03	28.26E05 (+)	24.18E05
AS29	64	−14.216	**−12.896**	1.229	−4.254	−3.372 (+)	0.875	62.73E04	25.72E05 (+)	21.51E05
AS30	64	−17.326	**−16.246**	0.980	−4.346	−3.256 (+)	0.952	26.95E03	30.08E05 (+)	28.34E05
AS31	64	−14.754	−12.953 (+)	1.576	−2.739	−2.179 (+)	0.486	85.24E03	47.35E05 (+)	54.02E05
AS32	64	−14.694	−14.340 (≈)	0.371	−3.387	−3.299 (+)	0.113	59.56E04	32.10E05 (+)	22.86E05
AS33	64	−13.751	−12.497 (≈)	1.132	−5.119	−3.713 (+)	1.231	10.25E05	24.44E05 (+)	12.34E05
AS34	64	−16.166	**−16.008**	0.177	−4.482	−3.936 (+)	0.483	80.52E01	46.96E05 (+)	40.01E05
AS35	64	−13.261	**−10.567**	2.357	−3.089	−2.785 (+)	0.373	57.57E04	48.62E05 (+)	45.54E05

Table 4.14 Results for six optimization algorithms on 30 real protein sequences using 2D AB off-lattice model

PS No.	PSL	GA			DE			PSO		
		Best	Mean	Std.	Best	Mean	Std.	Best	Mean	Std.
RS1	13	−2.269	−2.022 (≈)	0.168	−2.301	**−2.227**	0.073	−2.238	−2.039 (≈)	0.185
RS2	13	−3.485	−3.239 (≈)	0.296	−4.248	**−3.306**	0.608	−3.246	−2.552 (≈)	1.198
RS3	16	−7.582	−6.716 (≈)	0.734	−7.672	**−6.884**	0.608	−7.253	−6.449 (≈)	0.632
RS4	17	−6.088	−5.131 (≈)	0.632	−6.903	**−5.514**	0.862	−6.198	−5.015 (≈)	0.907
RS5	18	−6.197	−4.820 (+)	1.194	−6.095	−5.605 (≈)	0.339	−6.337	**−5.825**	0.726
RS6	21	−6.506	−5.093 (+)	1.338	−7.568	**−6.157**	0.864	−7.047	−5.883 (≈)	0.913
RS7	21	−6.224	−5.117 (≈)	1.243	−5.396	−5.143 (≈)	0.250	−6.726	**−5.922**	0.683
RS8	24	−16.405	−13.970 (+)	2.061	−13.517	−12.417 (+)	0.970	−19.041	**−16.956**	1.721
RS9	25	−6.027	−4.558 (≈)	1.653	−5.376	−4.605 (≈)	0.475	−7.049	**−5.129**	1.364
RS10	25	−11.372	−9.439 (+)	1.275	−9.122	−8.617 (+)	0.509	−13.497	**−11.211**	1.805
RS11	25	−11.202	−9.836 (≈)	1.071	−9.429	−9.181 (+)	0.180	−12.731	**−10.620**	1.280
RS12	26	−8.177	−6.582 (+)	1.111	−7.247	−6.733 (+)	0.297	−10.072	**−8.524**	1.315
RS13	29	−7.171	−4.753 (+)	1.464	−5.637	−5.263 (+)	0.324	−8.980	**−7.011**	1.565
RS14	29	−13.760	−12.063 (+)	1.255	−10.866	−9.336 (+)	0.922	−15.640	**−13.396**	2.077
RS15	29	−14.821	−10.057 (+)	3.039	−8.195	−7.964 (+)	0.203	−18.141	**−13.063**	4.117
RS16	29	−12.022	−9.704 (+)	3.110	−8.396	−7.978 (+)	0.347	−14.049	**−12.977**	1.205
RS17	29	−13.468	−9.068 (+)	2.852	−9.058	−8.046 (+)	0.863	−16.393	**−13.999**	2.158
RS18	34	−13.563	−12.099 (+)	2.299	−10.059	−9.156 (+)	0.839	−14.100	**−13.748**	1.435
RS19	37	−7.755	−7.027 (+)	0.658	−5.890	−5.256 (+)	0.514	−10.255	**−8.593**	1.712

(continued)

Table 4.14 (continued)

PS No.	PSL	GA			DE			PSO		
		Best	Mean	Std.	Best	Mean	Std.	Best	Mean	Std.
RS20	37	−11.383	−9.013 (+)	1.588	−8.477	−7.896 (+)	0.620	−16.312	**−12.935**	3.087
RS21	37	−9.448	−6.678 (+)	1.430	−6.223	−5.853 (+)	0.329	−14.096	**−10.196**	2.343
RS22	38	−15.207	−12.332 (+)	1.917	−10.070	−8.465 (+)	0.969	−18.296	**−15.732**	1.843
RS23	46	−27.604	−20.521 (+)	3.347	−13.223	−11.055 (+)	1.258	−27.874	**−24.256**	2.528
RS24	54	−14.978	−11.762 (+)	3.644	−8.106	−7.707 (+)	0.358	−19.696	**−18.019**	1.260
RS25	60	−13.676	−10.441 (+)	2.358	−5.349	−4.681 (+)	0.526	−18.196	**−17.125**	1.071
RS26	64	−13.885	−10.178 (+)	2.256	−3.808	−2.658 (+)	0.671	−16.149	**−15.980**	1.019
RS27	75	−13.129	−8.697 (+)	2.978	11.747	192.141 (+)	410.682	−16.575	**−14.253**	2.322
RS28	84	−14.655	−10.212 (+)	3.650	1.30E02	2.98E02 (+)	1.68E02	−23.178	**−21.928**	1.250
RS29	87	−19.259	−14.905 (+)	2.441	6.75E02	8.56E02 (+)	1.81E02	−19.307	−16.952 (+)	2.355
RS30	98	−34.801	−25.484 (+)	5.519	3.03E03	3.52E03 (+)	4.92E02	−25.523	−22.018 (+)	3.505

Table 4.14 (continued)

PS No.	PSL	ABC			CMA-ES			BA		
		Best	Mean	Std.	Best	Mean	Std.	Best	Mean	Std.
RS1	13	−1.961	−1.845 (≈)	0.128	−1.229	−1.119 (+)	0.112	−0.716	−0.299 (+)	0.454
RS2	13	−2.864	−2.464 (≈)	0.357	−0.889	−0.744 (+)	0.083	−0.532	0.108 (+)	0.554
RS3	16	−6.336	−6.080 (≈)	0.235	−2.865	−2.832 (+)	0.053	−3.543	−1.642 (+)	1.829
RS4	17	−3.699	−3.340 (+)	0.322	−1.627	−1.588 (+)	0.040	−1.516	0.215 (+)	1.518
RS5	18	−5.372	−5.005 (≈)	0.336	−2.388	−2.104 (+)	0.175	−0.656	0.287 (+)	0.977
RS6	21	−4.826	−4.686 (+)	0.137	−2.025	−1.942 (+)	0.055	0.894	3.700 (+)	2.535
RS7	21	−4.883	−4.759 (+)	0.191	−2.678	−2.556 (+)	0.124	−0.680	1.061 (+)	1.815
RS8	24	−16.170	−15.144 (+)	0.895	−7.050	−5.798 (+)	0.780	−2.513	1.914 (+)	3.876
RS9	25	−5.397	−5.076 (≈)	0.331	−2.141	−1.871 (+)	0.242	1.152	5.477 (+)	5.820
RS10	25	−9.348	−9.226 (+)	0.106	−3.136	−2.903 (+)	0.175	16.979	17.924 (+)	0.824
RS11	25	−10.193	−9.931 (≈)	0.251	−2.294	−2.115 (+)	0.110	0.847	6.760 (+)	5.727
RS12	26	−7.334	−7.101 (+)	0.202	−3.859	−3.361 (+)	0.337	0.274	28.616 (+)	33.781
RS13	29	−6.197	−5.881 (+)	0.283	−2.368	−2.193 (+)	0.116	59.231	83.695 (+)	29.619
RS14	29	−13.197	−12.201 (+)	0.976	−5.547	−4.880 (+)	0.459	12.76E01	37.46E01 (+)	34.73E01
RS15	29	−12.792	−12.179 (≈)	1.027	−4.104	−3.650 (+)	0.305	30.205	12.13E01 (+)	10.75E01
RS16	29	−11.936	−11.842 (≈)	0.142	−5.003	−4.093 (+)	0.731	5.716	88.833 (+)	73.954
RS17	29	−11.877	−10.930 (+)	0.930	−3.709	−3.195 (+)	0.424	4.660	43.21E01 (+)	63.88E01
RS18	34	−12.480	−12.324 (+)	0.211	−5.125	−4.529 (+)	0.513	10.967	43.90E02 (+)	62.35E02
RS19	37	−7.146	−6.982 (+)	0.193	−2.205	−1.956 (+)	0.275	13.05E02	17.29E03 (+)	22.25E03
RS20	37	−7.470	−5.927 (+)	1.206	−3.744	−1.986 (+)	0.999	37.46E01	31.12E03 (+)	30.02E03

(continued)

Table 4.14 (continued)

PS No.	PSL	ABC			CMA-ES			BA		
		Best	Mean	Std.	Best	Mean	Std.	Best	Mean	Std.
RS21	37	−9.990	−9.238 (≈)	1.196	−3.200	−2.797 (+)	0.334	52.94E01	22.63E02 (+)	27.19E02
RS22	38	−13.848	−13.247 (+)	0.521	−5.049	−4.301 (+)	0.375	12.277	16.04E03 (+)	23.58E03
RS23	46	−18.942	102.151 (+)	321.942	−7.819	−6.845 (+)	0.720	19.58E03	13.72E04 (+)	15.90E04
RS24	54	−15.904	−15.735 (+)	0.209	−4.400	−3.193 (+)	0.458	77.73E03	33.56E04 (+)	22.54E04
RS25	60	−16.483	−14.423 (+)	1.788	−4.898	−3.753 (+)	0.508	38.31E04	21.65E05 (+)	18.49E05
RS26	64	−15.207	−14.745 (+)	0.401	−5.091	−3.646 (+)	0.770	17.58E04	73.03E05 (+)	53.57E05
RS27	75	−13.268	−12.539 (+)	0.633	−3.923	−3.041 (+)	0.541	76.30E04	31.77E06 (+)	33.69E06
RS28	84	−21.151	−20.134 (+)	1.328	−3.195	−2.713 (+)	0.282	25.17E06	78.11E06 (+)	71.93E06
RS29	87	−25.784	**−24.845**	1.113	−5.224	−4.118 (+)	0.625	34.81E06	17.68E07 (+)	69.37E06
RS30	98	−46.547	**−44.992**	1.614	−8.124	−6.022 (+)	0.902	61.08E06	49.19E07 (+)	31.16E07

Table 4.15 Results for six optimization algorithms on 35 artificial protein sequences using 3D AB off-lattice model

PS No.	PSL	GA			DE			PSO		
		Best	Mean	Std.	Best	Mean	Std.	Best	Mean	Std.
AS1	13	−3.140	−2.614 (≈)	0.471	−1.966	−1.861 (+)	0.090	−1.408	−0.956 (+)	0.582
AS2	20	−9.712	−7.924 (+)	1.549	−1.609	−1.480 (+)	0.215	−5.156	−4.360 (+)	0.952
AS3	21	−6.472	−5.0587 (≈)	1.224	−0.680	−0.191 (+)	0.435	−4.424	−2.981 (+)	2.134
AS4	23	−7.746	−6.3889 (+)	1.324	0.629	1.592 (+)	0.909	−10.102	−4.574 (+)	4.848
AS5	24	−4.234	−3.470 (+)	0.667	2.893	3.282 (+)	0.455	0.506	3.237 (+)	3.381
AS6	25	−4.704	−3.403 (+)	1.148	3.830	4.316 (+)	0.452	0.206	4.012 (+)	4.150
AS7	34	−5.658	−4.996 (+)	1.078	5.495	6.653 (+)	1.031	−5.883	−1.782 (+)	6.360
AS8	36	−15.263	−11.411 (+)	3.477	7.158	7.749 (+)	0.529	−0.014	3.062 (+)	3.121
AS9	46	−21.270	−17.336 (+)	3.698	9.645	10.873 (+)	1.296	−1.680	1.557 (+)	4.530
AS10	48	−20.586	−19.788 (+)	0.694	11.242	12.853 (+)	1.472	6.619	8.401 (+)	2.463
AS11	48	−20.509	−12.581 (+)	7.011	14.638	15.599 (+)	0.899	−2.774	1.738 (+)	6.280
AS12	48	−17.244	−15.821 (+)	1.441	11.436	13.370 (+)	2.199	−17.173	−3.081 (+)	13.097
AS13	48	−15.815	−14.680 (+)	1.023	12.19	14.054 (+)	1.634	−13.314	−2.195 (+)	11.321
AS14	48	−13.424	−10.815 (+)	3.730	12.863	13.888 (+)	0.912	−10.502	−2.665 (+)	7.841
AS15	48	−17.130	−15.092 (+)	2.889	12.807	13.887 (+)	1.258	−17.276	−5.003 (+)	11.937
AS16	48	−12.929	−11.639 (+)	1.120	7.492	10.607 (+)	2.698	−21.494	−3.162 (+)	16.502
AS17	48	−18.919	−17.732 (+)	1.964	15.346	15.684 (+)	0.293	−28.551	−15.872 (+)	11.019
AS18	48	−18.667	−15.573 (+)	2.755	12.819	13.885 (+)	1.602	−6.634	0.145 (+)	8.050
AS19	48	−15.608	−10.714 (+)	5.456	11.511	12.232 (+)	0.681	−4.974	−3.066 (+)	2.324

(continued)

Table 4.15 (continued)

PS No.	PSL	GA			DE			PSO		
		Best	Mean	Std.	Best	Mean	Std.	Best	Mean	Std.
AS20	48	−14.279	−9.731 (+)	4.219	13.721	15.157 (+)	1.246	−24.292	−10.851 (+)	12.472
AS21	50	−16.088	−14.624 (+)	1.492	10.914	12.065 (+)	0.998	−9.279	−4.219 (+)	5.149
AS22	55	−7.403	−5.220 (+)	2.099	17.963	20.835 (+)	4.434	−4.596	3.109 (+)	6.741
AS23	58	−19.019	−15.341 (+)	4.048	23.847	26.915 (+)	3.257	−5.495	1.129 (+)	5.737
AS24	60	−40.216	−34.850 (+)	5.857	9.348	13.608 (+)	3.744	−36.429	−26.925 (+)	9.579
AS25	64	−29.353	−27.700 (+)	1.858	19.716	29.893 (+)	8.815	−12.661	−6.132 (+)	3.905
AS26	64	−3.599	−0.994 (+)	3.256	28.509	43.314 (+)	22.020	8.822	15.243 (+)	4.008
AS27	64	−3.642	−2.077 (+)	2.545	28.531	31.061 (+)	4.258	−10.041	1.323 (+)	7.084
AS28	64	−11.452	−10.196 (+)	1.402	29.069	34.880 (+)	6.753	−22.995	1.677 (+)	14.697
AS29	64	−5.547	−1.954 (+)	3.167	27.232	39.111 (+)	9.711	−3.254	5.750 (+)	11.190
AS30	64	−10.201	−5.855 (+)	3.942	32.292	38.479 (+)	8.139	1.461	5.682 (+)	3.594
AS31	64	−4.568	−1.625 (+)	2.332	25.153	46.008 (+)	12.395	9.355	15.731 (+)	4.428
AS32	64	−9.844	−6.071 (+)	2.851	24.576	36.554 (+)	8.146	−7.342	2.794 (+)	6.387
AS33	64	−11.991	−4.574 (+)	4.030	31.740	45.211 (+)	10.708	−0.120	10.027 (+)	6.565
AS34	64	−12.651	−7.226 (+)	2.998	28.225	39.622 (+)	7.296	−22.561	6.044 (+)	10.815
AS35	64	−6.675	−2.392 (+)	2.755	28.398	39.435 (+)	7.569	−10.864	9.826 (+)	8.675

Table 4.15 (continued)

PS No.	PSL	ABC			CMA-ES			BA		
		Best	Mean	Std.	Best	Mean	Std.	Best	Mean	Std.
AS1	13	−3.377	**−2.855**	0.320	−1.425	−1.200 (+)	0.174	−0.516	0.699 (+)	0.485
AS2	20	−11.314	**−9.357**	1.122	−3.858	−2.999 (+)	0.508	−0.418	1.830 (+)	1.180
AS3	21	−5.751	**−5.507**	0.311	−3.068	−2.666 (+)	0.285	1.159	2.136 (+)	0.859
AS4	23	−10.074	**−9.252**	0.762	−2.804	−2.756 (+)	0.063	2.102	4.084 (+)	1.718
AS5	24	−6.666	**−5.851**	0.888	−0.369	−0.187 (+)	0.174	3.474	5.107 (+)	1.417
AS6	25	−7.969	**−7.339**	0.981	−0.301	−0.223 (+)	0.068	4.959	6.067 (+)	1.016
AS7	34	−9.306	**−9.231**	0.074	−3.595	−3.056 (+)	0.715	6.294	27.085 (+)	25.570
AS8	36	−20.953	**−19.367**	2.183	−5.760	−4.919 (+)	0.751	7.320	29.941 (+)	19.828
AS9	46	−38.158	**−37.111**	1.322	−4.154	−3.882 (+)	0.241	13.683	35.56E01 (+)	44.84E01
AS10	48	−32.214	**−31.883**	0.400	−5.973	−5.750 (+)	0.312	14.089	24.30 (+)	21.50E01
AS11	48	−36.948	**−31.076**	5.277	−3.733	−3.581 (+)	0.140	12.4056	74.464 (+)	59.0585
AS12	48	−29.065	**−27.734**	1.319	−3.075	−2.966 (+)	0.094	12.085	30.34 E01 (+)	37.96E01
AS13	48	−27.392	**−26.071**	2.060	−4.653	−4.232 (+)	0.467	9.5045	15.55E01 (+)	13.28E01
AS14	48	−27.805	**−26.823**	1.308	−4.073	−3.770 (+)	0.487	12.851	51.59E01 (+)	46.77E01
AS15	48	−31.312	**−30.062**	1.112	−4.202	−3.677 (+)	0.610	13.731	17.75E01 (+)	16.13E01
AS16	48	−28.985	**−28.721**	0.382	−6.828	−5.435 (+)	0.843	12.324	28.35E01 (+)	36.48E01
AS17	48	−31.238	**−29.693**	1.707	−3.985	−3.096 (+)	0.516	15.416	61.27E01 (+)	78.88E01
AS18	48	−35.763	**−31.813**	3.605	−3.256	−2.873 (+)	0.386	13.813	16.78E01 (+)	15.93E01
AS19	48	−31.437	**−30.602**	1.054	−5.878	−5.051 (+)	0.514	11.669	34.34E01 (+)	28.73E01

(continued)

Table 4.15 (continued)

PS No.	PSL	ABC			CMA-ES			BA		
		Best	Mean	Std.	Best	Mean	Std.	Best	Mean	Std.
AS20	48	−29.780	−26.349	3.470	−3.8078	−3.506 (+)	0.269	14.845	16.97E01 (+)	17.83E01
AS21	50	−33.415	−31.909	1.329	−6.083	−5.007 (+)	0.784	10.1534	36.20E01 (+)	44.18E01
AS22	55	−16.239	−15.513	1.009	−4.396	−3.792 (+)	0.383	18.169	77.78E01 (+)	88.29E01
AS23	58	−34.880	−33.717	0.795	−2.779	−2.364 (+)	0.298	14.982	81.33E01 (+)	12.14E02
AS24	60	−84.939	−78.941	4.336	−13.536	−12.080 (+)	1.058	6.7805	39.78E02 (+)	40.45E02
AS25	64	−−62.977	−59.575	2.857	−11.216	−9.891 (+)	0.927	13.104	35.34E02 (+)	44.46E02
AS26	64	−21.387	−19.935	1.399	−2.3766	−2.130 (+)	0.302	25.431	46.71E02 (+)	42.63E02
AS27	64	−29.270	−26.889	1.687	−4.341	−3.412 (+)	0.609	18.954	32.84E02 (+)	41.12E02
AS28	64	−42.390	−40.358	1.559	−6.444	−5.629 (+)	0.521	18.434	47.71E02 (+)	41.16E02
AS29	64	−25.590	−23.505	1.311	−2.820	−2.535 (+)	0.223	23.309	20.14E02 (+)	17.29E02
AS30	64	−33.784	−30.332	2.509	−2.716	−2.459 (+)	0.221	22.534	51.73E02 (+)	38.19E02
AS31	64	−27.860	−23.932	1.675	−2.645	−2.187 (+)	0.255	21.735	55.51E02 (+)	50.32E02
AS32	64	−32.320	−27.364	2.519	−4.307	−3.671 (+)	0.369	21.915	42.67E02 (+)	27.71E02
AS33	64	−33.553	−28.468	2.675	−4.476	−3.165 (+)	0.543	21.460	54.55E02 (+)	44.23E02
AS34	64	−29.530	−27.509	1.477	−5.144	−3.663 (+)	0.651	22.230	54.97E02 (+)	35.54E02
AS35	64	−21.008	−18.782	1.568	−2.995	−2.074 (+)	0.439	21.956	50.55E02 (+)	33.37E02

Table 4.16 Results for six optimization algorithms on 30 real protein sequences using 3D AB off-lattice model

PS No.	PSL	GA			DE			PSO		
		Best	Mean	Std.	Best	Mean	Std.	Best	Mean	Std.
RS1	13	−3.811	**−3.307**	0.392	−2.650	−2.333 (+)	0.189	−4.015	−2.305 (+)	0.889
RS2	13	−6.520	−4.450 (≈)	1.542	−3.164	−2.524 (+)	0.489	−5.580	−3.321 (+)	1.619
RS3	16	−12.222	−10.148 (≈)	1.540	−5.538	−4.457 (+)	0.770	−11.112	−6.338 (+)	3.774
RS4	17	−9.358	−4.734 (+)	1.908	−1.437	−0.826 (+)	0.394	−9.528	**−7.701**	2.005
RS5	18	−11.830	**−8.919**	2.052	−2.143	−1.407 (+)	0.492	−9.862	−3.788 (+)	3.068
RS6	21	−10.035	−6.859 (+)	1.379	−0.536	0.389 (+)	0.616	−10.357	−4.041 (+)	3.095
RS7	21	−8.515	−6.563 (+)	1.577	−1.725	−0.746 (+)	0.691	−7.537	−2.801 (+)	3.341
RS8	24	−27.489	−23.548 (+)	2.027	−9.568	−7.297 (+)	1.256	−32.579	−16.176 (+)	8.696
RS9	25	−6.473	−5.222 (+)	0.795	1.471	1.992 (+)	0.394	−8.388	−1.034 (+)	4.164
RS10	25	−13.438	−11.367 (+)	1.807	−1.246	0.120 (+)	0.941	−17.888	−5.514 (+)	5.973
RS11	25	−17.104	−12.354 (+)	4.110	−2.530	−0.343 (+)	1.105	−15.239	−5.967 (+)	5.622
RS12	26	−11.044	−10.130 (≈)	1.208	0.632	0.976 (+)	0.300	−6.251	−2.863 (+)	5.331
RS13	29	−9.114	−6.822 (+)	2.158	1.777	2.821 (+)	0.915	−7.223	−3.555 (+)	4.020
RS14	29	−16.273	−15.576 (+)	0.841	−2.094	−0.724 (+)	1.350	−12.663	−7.996 (+)	7.395
RS15	29	−20.159	−16.794 (+)	3.269	2.423	2.776 (+)	0.460	−14.137	−7.705 (+)	9.608
RS16	29	−17.097	−16.041 (+)	1.386	−0.321	0.861 (+)	1.314	−4.260	−1.894 (+)	2.300
RS17	29	−22.702	−16.065 (+)	5.965	2.373	2.680 (+)	0.278	−17.527	−2.726 (+)	12.818
RS18	34	−19.416	−15.265 (+)	4.485	3.352	4.564 (+)	1.086	−14.386	−4.878 (+)	9.076
RS19	37	−10.500	−8.674 (+)	1.674	9.786	10.299 (+)	0.553	−4.193	−2.690 (+)	1.938

(continued)

Table 4.16 (continued)

PS No.	PSL	GA				DE				PSO		
		Best	Mean	Std.		Best	Mean	Std.		Best	Mean	Std.
RS20	37	−12.256	−10.836 (+)	1.255		7.926	8.419 (+)	0.437		−1.522	2.728 (+)	3.700
RS21	37	−11.667	−9.743 (+)	2.141		9.203	9.509 (+)	0.494		−2.993	3.970 (+)	6.071
RS22	38	−22.168	−17.337 (+)	4.229		6.724	6.943 (+)	0.190		−7.818	−3.712 (+)	3.819
RS23	46	−28.426	−24.720 (+)	5.575		7.381	8.925 (+)	1.411		−23.621	−15.099 (+)	10.751
RS24	54	−17.882	−14.035 (+)	4.135		18.640	20.943 (+)	2.022		−9.799	2.247 (+)	10.439
RS25	60	−7.578	−3.273 (+)	5.189		29.447	31.377 (+)	2.753		5.294	10.579 (+)	5.218
RS26	64	−7.210	−2.510 (+)	4.140		31.243	35.246 (+)	6.000		−6.415	8.140 (+)	12.729
RS27	75	−0.363	1.891 (+)	2.122		67.999	97.616 (+)	49.916		−17.798	−9.077 (+)	12.990
RS28	84	2.541	3.431 (+)	1.000		114.614	483.081 (+)	504.229		17.179	21.368 (+)	4.432
RS29	87	0.039	1.767 (+)	2.651		205.482	427.891 (+)	202.463		−8.623	10.466 (+)	18.399
RS30	98	−13.704	−6.738 (+)	7.101		1673.754	2273.854 (+)	912.044		−43.612	−14.893 (+)	25.103

Table 4.16 (continued)

PS No.	PSL	ABC			CMA-ES			BA		
		Best	Mean	Std.	Best	Mean	Std.	Best	Mean	Std.
RS1	13	−3.396	−2.948 (≈)	0.229	−0.954	−0.725 (+)	0.098	−0.430	0.286 (+)	0.308
RS2	13	−6.293	−4.739	0.799	−1.240	−0.887 (+)	0.129	−0.688	0.991 (+)	0.620
RS3	16	−11.561	−10.197	0.865	−4.514	−3.495 (+)	0.408	−2.668	−0.250 (+)	1.197
RS4	17	−7.273	−6.344 (+)	0.619	−1.900	−1.510 (+)	0.148	1.202	2.442 (+)	0.494
RS5	18	−10.447	−8.383 (≈)	1.111	−2.571	−2.318 (+)	0.107	−0.765	1.517 (+)	0.990
RS6	21	−11.909	−9.398	1.283	−2.384	−2.127 (+)	0.143	0.90	3.899 (+)	1.210
RS7	21	−10.170	−9.018	0.726	−2.338	−2.212 (+)	0.104	1.972	3.363 (+)	0.755
RS8	24	−27.702	−26.346	0.901	−8.560	−7.203 (+)	0.696	−6.262	−0.367 (+)	2.935
RS9	25	−8.920	−8.098	0.761	−1.545	−1.397 (+)	0.138	1.839	5.668 (+)	2.238
RS10	25	−18.698	−15.077	1.487	−3.552	−3.032 (+)	0.306	0.374	5.482 (+)	2.271
RS11	25	−17.413	−15.915	1.429	−3.436	−2.943 (+)	0.371	−0.276	4.122 (+)	2.874
RS12	26	−12.567	−10.955	1.010	−4.216	−3.851 (+)	0.228	1.562	4.064 (+)	1.567
RS13	29	−12.727	−10.707	1.554	−2.774	−2.385 (+)	0.268	4.289	7.354 (+)	4.028
RS14	29	−24.248	−23.063	0.996	−5.334	−4.749 (+)	0.460	0.073	6.421 (+)	5.565
RS15	29	−23.374	−20.718	1.925	−4.050	−3.320 (+)	0.480	4.421	6.969 (+)	2.550
RS16	29	−23.371	−21.735	1.607	−3.722	−3.332 (+)	0.312	3.947	10.428 (+)	6.132
RS17	29	−21.480	−18.946	1.644	−3.315	−2.916 (+)	0.319	4.263	8.074 (+)	4.710
RS18	34	−20.971	−20.197	0.656	−4.548	−4.194 (+)	0.319	3.225	13.814 (+)	9.314
RS19	37	−16.078	−13.839	1.433	−2.320	−1.704 (+)	0.384	11.885	24.930 (+)	14.910

(continued)

Table 4.16 (continued)

PS No.	PSL	ABC			CMA-ES			BA		
		Best	Mean	Std.	Best	Mean	Std.	Best	Mean	Std.
RS20	37	−21.259	**−17.779**	2.158	−2.962	−2.366 (+)	0.426	9.979	39.858 (+)	28.290
RS21	37	−16.090	**−15.406**	0.622	−2.908	−2.703 (+)	0.178	9.010	32.392 (+)	27.269
RS22	38	−23.939	**−23.279**	0.947	−4.641	−4.333 (+)	0.320	9.022	39.508 (+)	26.481
RS23	46	−45.087	**−42.661**	2.224	−7.107	−6.540 (+)	0.980	8.510	164.365 (+)	135.317
RS24	54	−28.068	**−28.003**	0.083	−3.558	−3.310 (+)	0.260	16.273	11.30E02 (+)	96.58E01
RS25	60	−29.738	**−27.689**	2.256	−2.933	−2.754 (+)	0.155	19.629	53.34E02 (+)	53.43E02
RS26	64	−27.791	**−24.381**	3.009	−4.802	−3.298 (+)	1.305	22.918	33.24E02 (+)	43.24E02
RS27	75	−28.213	**−26.433**	1.570	−2.422	−2.289 (+)	0.119	29.731	66.68E02 (+)	58.42E02
RS28	84	−35.442	**−33.698**	2.624	−3.213	−2.886 (+)	0.288	33.458	27.99E03 (+)	37.15E03
RS29	87	−52.407	**−49.459**	3.269	−3.688	−3.367 (+)	0.305	33.496	28.49E03 (+)	30.13E03
RS30	98	−93.132	**−90.095**	3.485	−7.107	−6.537 (+)	0.675	26.037	10.20E04 (+)	90.11E03

with almost same *mean* results for the sequences AS2, AS5, AS13, RS1, RS3, and RS4. However, PSO achieves worst *mean* results on AS35, RS29, and RS30. The *standard deviation* results of PSO indicate that the best results are highly robust in each run for all the protein instances. Overall, the performances of the PSO algorithm are statistically significant than the other algorithms. Balanced exploration and exploitation by PSO play a crucial role in overcoming the difficulties are associated with the landscape structure of the PSP problem.

The ABC algorithm performs well for the protein sequences AS13, AS22, AS23, AS25, AS27, AS29, AS30, and AS34. Also, exceptionally good results are obtained on the real protein sequences RS29 and RS30 by ABC. ABC offered superior performance on the higher lengths of protein sequences AS35, RS29, and RS30 whose structural features are very sensitive compared to smaller lengths of protein sequences. The ABC algorithm has been empowered with employed and onlooker bees which can successfully explore the rugged multi-funnel landscape structure and avoid the local optima stagnation in the case of higher lengths of protein sequences.

Focusing the results of CMA-ES in Tables 4.13 and 4.14, it is observed that CMA-ES offers worst *mean* results compared to the GA, DE, PSO and ABC algorithms on most of the protein instances. CMA-ES dominates to BA on for each protein sequence. CMA-ES appears to yield better results than the DE and PSO algorithm for RS26-RS30 but not providing best results. Thus, the standard version of the CMA-ES algorithm may not be suitable for solving the PSP problem. However, the robustness of the results is promising on real protein sequences.

We observed worst performance in BA for each protein sequence. Unexpected poor results are achieved on higher lengths of protein sequences indicates that solutions are probably stuck at local optima. Hence, the search space is not properly explored with the mentioned configuration of BA. Neighborhood structure around the scout bees is responsible for exploring the search space and configured neighborhood used in this study may be a bad one. Although, the performance of BA can be improved with the proper selection of neighborhood structure or local search [94]. Also, selection of best scout bees plays an important role in BA to improve performance.

Convergence characteristics of the algorithms have been shown for some artificial and real protein instances in Figs. 4.24 and 4.25. The plots correspond to the best run of each algorithm. The plots are revealed that the PSO algorithm converges faster among the other algorithms on the majority of the protein instances. Exceptionally, the ABC algorithm could reach faster to minimum energy value for the higher lengths protein sequences.

4.5.3.2   For 3D AB Off-Lattice Model

Tables 4.15 and 4.16 shows the *best*, *mean* and *standard deviation (std.)* of the energy values achieved by the algorithms on ASs and RSs using 3D AB off-lattice model. In 3D AB off-lattice model, an n length protein sequence is represented by $(n - 2)$ bond angles and $(n - 3)$ torsion angles. Thus, a solution can be encoded as a $(2n - 5)$

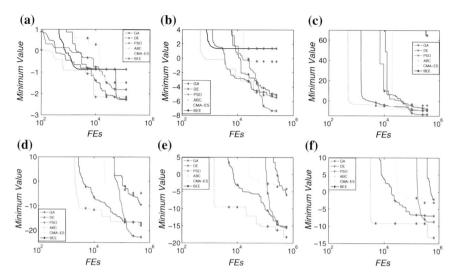

Fig. 4.24 Convergence curve of algorithms for ASs in 2D AB off-lattice model: **a** AS1 with length 13, **b** AS6 with length 25, **c** AS8 with length 36, **d** AS15 with length 48, **e** AS23 with length 58 and **f** AS35 with length 64

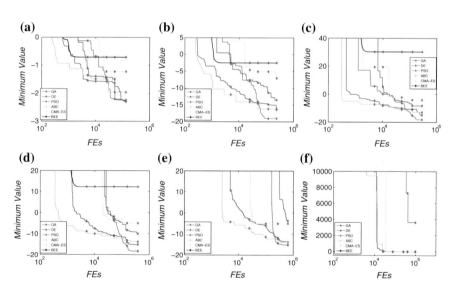

Fig. 4.25 Convergence curve of algorithms for RSs in 2D AB off-lattice model: **a** RS1 with length 13, **b** RS8 with length 24, **c** RS15 with length 29, **d** RS22 with length 38, **e** RS26 with length 64 and **f** RS30 with length 98

dimensional vector while a solution in 2D AB off-lattice model encoded as only $(n - 2)$ dimensional vector. Therefore, the PSP problem in 3D AB off-lattice model inherently include high dimensionality feature other than the features discussed in Sect. 4.4.2 for each protein sequence.

GA achieved well on the *best* results for the sequences AS3, RS2, RS3, RS5 and RS17 out of the 65 protein sequences. On RS1 and RS5, the best *mean* result is obtained by GA. Also, GA has comparable performance with the *mean* results obtained by the best algorithm for the sequence AS1, AS3, RS2, RS3, and RS12. Overall, performances of the GA algorithm are not promising for solving the PSP problem and statistical not significant.

The DE algorithm performs poorly for each protein sequences in the 3D AB off-lattice model. In this model dimensions of a candidate, solution is approximately double for each protein sequences and hence complexity increases compared to 2D AB off-lattice model. The standard configurations of the DE algorithm are unable to efficiently explore the search landscape and possibly the solutions are stuck at local optima provides poor results. The PSO algorithm achieved best on *best* results for the sequences AS4, RS1, RS4 and RS8 compared to the other algorithms. On RS4 best *mean* result is obtained by PSO. PSO does not perform well on the protein sequences which are defined on 3D AB off-lattice model. It has been verified by the statistical test that the PSO algorithm does not possess statistical significance on the *mean* with the ABC algorithms.

An outstanding performance has been achieved by the ABC algorithm. The *mean* results are obtained by ABC dominates all other algorithms for the protein instances irrespective of sequence lengths expecting smaller lengths sequences RS1, RS4, and RS5. It performs well on the smaller as well as higher lengths of sequences regardless type of the protein sequences. A significant result has been observed on the higher lengths of protein sequences by the ABC algorithm. Overall, the performance of the ABC algorithm is statistically significant on *mean* results compared to the other algorithms for all the protein sequences. Hence, ABC can be the most appropriate algorithm to find the minimum energy value for the PSP problem in 3D AB off-lattice model. The ABC algorithm can steadily make their progress towards the minimum energy value avoiding trapped at local optima.

Worst results are obtained for the protein sequences by CMA-ES compared to the ABC algorithm but dominate the *mean* results are obtained by using the PSO, DE and BA algorithms on most of the protein sequences. In the limited number of function evaluations, the obtained results are appreciable and expected to get promising results after increasing the function evaluations well as the incorporation of the new operator in the framework of the CMA-ES algorithm to enhanced the exploration capability. Therefore, PSP problem can be solved by the CMA-ES algorithm with proper configurations. The worst performance is obtained by BA for solving the protein sequence in the 3D AB off-lattice model. The achieved results are not up to the mark. An extensive investigation can be required on the configurations of the BA for solving the PSP problem in 3D AB off-lattice model.

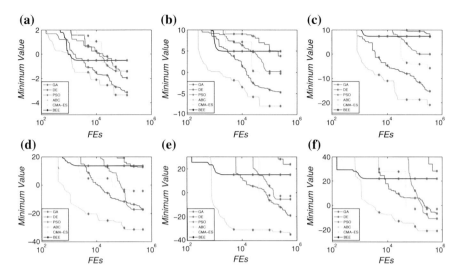

Fig. 4.26 Convergence curve of algorithms for ASs in 3D AB off-lattice model: **a** AS1 with length 13, **b** AS6 with length 25, **c** AS8 with length 36, **d** AS15 with length 48, **e** AS23 with length 58 and **f** AS35 with length 64

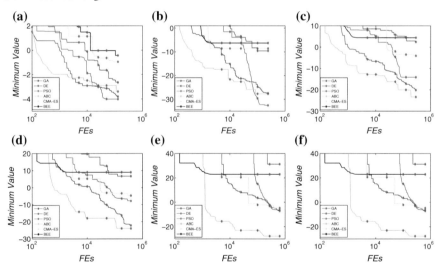

Fig. 4.27 Convergence curve of algorithms for RSs in 3D AB off-lattice model: **a** RS1 with length 13, **b** RS8 with length 24, **c** RS15 with length 29, **d** RS22 with length 38, **e** RS26 with length 64 and **f** RS30 with length 98

Figures 4.26 and 4.27 depicts the convergence characteristics for all the algorithms on selected ASs and RSs taken from each group. Convergence plots provided faster convergence of the ABC algorithms compared to the other algorithms.

Table 4.17 Summary of landscape features and most appropriate algorithm for protein sequences in 2D AB off-lattice model

Model	Group	Landscape features	Algorithms
2D	G_{A_1}	Landscape structure of each ASs are highly rugged, multi-modal, deceptive, Presence of many needle-like funnels. Also, the number of local optima for the protein sequences increase as sequence length increase.	1st: PSO 2nd: GA
	G_{A_2}	Landscape structure of each ASs are highly rugged, multi-modal, deceptive, Presence of many needle like funnels. Number of local optima increases as sequence length increase compared to G_{A_1}.	1st: PSO 2nd: ABC
	G_{A_3}	Landscape structure of each ASs highly rugged, multi-modal, deceptive, Presence of many needle like funnels. Number of local optima on landscape structure is more compared to G_{A_2} but approximately same for all the AS's due to same length of protein sequences in this group.	1st: PSO 2nd: ABC
	G_{A_4}	Landscape structure of each ASs in this group are highly rugged, multi-modal, deceptive, Presence of many needle like funnels. Number of local optima for the protein sequences are increases as sequence length increase.	1st: PSO 2nd: ABC
	G_{A_5}	Landscape structure of each ASs highly rugged, multi-modal, deceptive, Presence of many needle like funnels. Number of local optima on landscape structure is more compared to G_{A_4} but approximately same for all the AS's due to same length of protein sequences in this group.	1st: ABC 2nd: PSO
	G_{R_1}	Landscape structure of each ASs in this group are highly rugged, multi-modal, deceptive, Presence of many needle like funnels. Also, number of local optima for the protein sequences are increases as sequence length increase.	1st: DE 2nd: PSO
	G_{R_2}	Landscape structure of each ASs are highly rugged, multi-modal, deceptive, Presence of many needle like funnels. Number of local optima for the protein sequences are increases as sequence length increase.	1st: PSO 2nd: ABC
	G_{R_3}	Landscape structure of each ASs in the group are highly rugged, multi-modal, deceptive, consisting many needle like funnels. Local optima on landscape structure is more compared to G_{R_2} but approximately same for all the RS's due to same length of protein sequences in this group.	1st: PSO 2nd: ABC
	G_{R_4}	Landscape structure of each ASs are highly rugged, multi-modal, deceptive, Presence of many needle like funnels. Number of local optima is increases as sequence length increase.	1st: PSO 2nd: ABC
	G_{R_5}	Landscape structure of each ASs are highly rugged, multi-modal, deceptive, presence of many needle like funnels. Significant change in number of local optima occurred for the protein sequences are increases as sequence length increase.	1st: PSO 2nd: ABC

Table 4.18 Summary of landscape features and most appropriate algorithm for protein sequences in 3D AB off-lattice model

Model	Group	Landscape features	Algorithms
3D	G_{A_1}	Landscape structure for all protein sequences in this group are highly rugged, multi-modal, deceptive, consisting many needle like funnels. Local optima increases as sequence length increase. Dimensions increases as sequence length increases are compared to 2D model.	1st: ABC 2nd: GA
	G_{A_2}	Landscape structure for all protein sequences in this group are highly rugged, multi-modal, deceptive, a presence of many needle-like funnels. The number of local optima increases as sequence length increase. Also, dimensions in encoded solution increase as sequence length increase.	1st: ABC 2nd: GA
	G_{A_3}	Landscape structure for all protein sequences are highly rugged, multi-modal, deceptive, a presence of many needle-like funnels. The number of local optima approximately same for each sequence. Also, dimensions in encoded solution increase as sequence length increase.	1st: ABC 2nd: GA
	G_{A_4}	Landscape structure for all protein sequences are highly rugged, multi-modal, deceptive, a presence of many needle-like funnels. The number of local optima increases as sequence length increase. Also, dimensions in encoded solution increase as sequence length increase.	1st: ABC 2nd: GA
	G_{A_5}	Landscape structure for all protein sequences are highly rugged, multi-modal, deceptive, a presence of many needle-like funnels. The number of local optima approximately same for each sequence. Also, dimensions in encoded solution increase as sequence length increase.	1st: ABC 2nd: GA
	G_{R_1}	Landscape structure for all protein sequences in this group are highly rugged, multi-modal, deceptive, a presence of many needle-like funnels. The number of local optima increases as sequence length increase. Also, dimensions in encoded solution increase as sequence length increase.	1st: ABC 2nd: GA
	G_{R_2}	Landscape structure for all protein sequences in this group are highly rugged, multi-modal, deceptive, a presence of many needle-like funnels. The number of local optima increases as sequence length increase. Also, dimensions in encoded solution increase as sequence length increase.	1st: ABC 2nd: GA
	G_{R_3}	Landscape structure for all protein sequences are highly rugged, multi-modal, deceptive, a presence of many needle-like funnels. The number of local optima approximately same for each sequence in this group. Also, dimensions in encoded solution increase as sequence length increase.	1st: ABC 2nd: GA
	G_{R_4}	Landscape structure for all protein sequences are highly rugged, multi-modal, deceptive, a presence of many needle-like funnels. The number of local optima increases as sequence length increase. Also, dimensions in encoded solution increase as sequence length increase.	1st: ABC 2nd: GA
	G_{R_5}	Landscape structure for all protein sequences are highly rugged, multi-modal, deceptive, a presence of many needle-like funnels. The number of local optima approximately same as sequence length increases. Also, dimensions in encoded solution increase as sequence length increase.	1st: ABC 2nd: GA

From the above results and discussions, we summarize the landscape features and the most appropriate algorithm for solving the PSP problem in both the 2D and 3D AB off-lattice model are given in Tables 4.17 and 4.18. All the protein instances in both the models are highly rugged, contains multiple local optima, deceptive and consists many funnels. Although, the number of local optima and funnels are increased as sequence length increases in both the models. Moreover, one more feature, high dimensionality is associated in the 3D AB off-lattice model may become difficult to solve the PSP problem compared to the 2D AB off-lattice model. Most appropriate algorithms are ranked based on *mean* results to the corresponding group of the protein instances in Tables 4.17 and 4.18. Finally, the PSO and ABC algorithms show significant improvement while solving the protein instances in 2D AB off-lattice model. The ABC and GA algorithms are most promising algorithms to solve the protein instances in 3D AB off-lattice model. However, modifications or improvements can be expected on the operators used in the algorithms to achieve the better results on protein energy value.

4.6 Summary

This chapter introduced quasi-random sampling technique and city block distance to generate landscape-path for the proteins instances. We determine the structural features of the PSP problem using FLA techniques based on the generated landscape path. The 65 protein sequences of both artificial and real protein instances are considered in the experiments on 2D and 3D AB off-lattice models and analyzed the performance of the six widely used optimization algorithms using structural features.

From the results of FLA, it has been shown that the PSP problem has high rugged landscape structure, contains many local optima and needle-like funnel with no global structure that characterize the problem complexity. In addition, experimental results reveal that problem complexity increase as protein sequence length increases in both models irrespective of the type of the protein instances. Pearsons correlation coefficient demonstrates that most of the landscape measures are not correlated to each other in both 2D and 3D AB off-lattice model. Experimental results show that the PSO algorithm outperforms on most of the protein instances compared to other algorithms in 2D AB off-lattice model. For few higher length artificial and real protein sequences, PSO does not perform well while ABC shows superior performance on that protein sequences. PSO and ABC algorithm has potential ability to explore the search space and hence provides better performance. On the other hand, experiment and analysis also show that the ABC algorithm significantly outperforms on all the protein instances compared to other algorithms in 3D AB off-lattice model. Also, GA shows good results on few sequences compared to PSO, DE, CMA-ES, and BA. Therefore, PSO and ABC algorithms are most appropriate for solving the PSP problem in 2D AB off-lattice model while ABC and GA algorithms are effective and promising on 3D AB off-lattice model.

Chapter 5
The Lévy Distributed Parameter Adaptive Differential Evolution for Protein Structure Prediction

Abstract This chapter introduces a scheme for controlling parameters adaptively in the differential evolution (DE) algorithm. In the proposed method, parameters of DE are adapted using Lévy distribution. The distribution function allows possible changes with a significant amount in the control parameters adaptively, which provides good exploration and exploitation in the search space to reach the global optimum point. The performance of the Lévy distributed DE algorithm has been extensively studied and compared with some parameter control techniques over a test-suite of unimodal, basic and expanded multi-modal and hybrid composite functions with different dimensions. The proposed method is also investigated on real protein sequences for protein structure prediction. The results exhibits that the Lévy distributed DE algorithm provides significant performance in terms of accuracy and convergence speed to obtain global optimum solution.

5.1 Introduction

The protein structure prediction (PSP) using computational technique is *NP*-hard and a non-linear complex optimization problem. Moreover, the problem has some internal characteristic (as discussed in Chap. 4) that makes PSP is a highly multi-modal, rugged and consisting many funnels without global structure. Such characteristics influence the search performance of a metaheuristic algorithm. So, proper configuration or parameter control of an algorithm is needed to efficiently search the global optimum on the protein energy function landscape, otherwise, a premature solution would reach due to local optima. The performance of an algorithm largely depends on the trade-off between exploration and exploitation dilemma during the search process which is taken care off by the control of the parameters.

© Springer International Publishing AG 2018

N. D. Jana et al., *A Metaheuristic Approach to Protein Structure Prediction*, Emergence, Complexity and Computation 31,
https://doi.org/10.1007/978-3-319-74775-0_5

The DE algorithm is a population-based metaheuristic search algorithm proposed by Storn and Price [173], has been discussed in Chap. 1. The algorithm is very simple, robust and having faster convergence speed to find the global optimal solution over the continuous search space. However, DE produced premature solutions when solving PSP problem or multi-modal problems. Basically, a performance of the DE algorithm depends on proper adjustment between exploration and exploitation strategies during the execution processes [174]. If the balance between exploration and exploitation strategy is hampered, the problem like stagnation of the population, premature convergence may appear and the situation occurs when solving PSP problem. The limitations of DE can be handled with the parameter control mechanism [146]. In the mechanism, parameters of an algorithm are controlled in every generation or in individual level in deterministic [175], adaptive [176] and self-adaptive [64] way. The DE algorithm consists three crucial parameters, population size (NP), scaling factor (F) and crossover ratio (CR). The parameters significantly influence the performance of the DE algorithm when solving a problem. In the evolution processes, DE explores different regions in the search space by selecting different values for the specific parameter. Therefore, it is required to adaptively determine the values of the associated parameters at different stages of the evolution process.

Researchers have developed a number of modified versions of DE which are free from manual settings of the control parameters. Abbas et al. [177] introduced a self-adaptive approach to obtain DE parameters using variable step length, generated by a Gaussian distribution which is evolved during the optimization process. Fuzzy adaptive differential evolution algorithm based on fuzzy logic controllers was proposed in [178]. In the method, fuzzy inputs incorporate the relative function values and individuals to adapt the parameters for the mutation and crossover operation at successive generations. Das et al. [175] employed two schemes for adapting the scale factor F in DE. In the first scheme, the scale factor varied randomly between 0.5 and 1.0 in successive generations. Scale factor F linearly decrease from 1.0 to 0.5 in the second scheme. The proposed approach encourage individuals to explore diverse regions of the search space during the early stages of execution. In the later execution stage, a decaying scale factor helps to adjust the movements of trial solutions, so that the selected individuals explore the interior of a relatively small region of the search space. Brest et al. [64] proposed self-adaptive mechanisms to control the parameters F and CR of DE by encoding into individual and adjusted with the help of two new parameters τ_1 and τ_2. A set of F and CR values are assigned to each individual in the population for augmenting the dimensions of each vector in the proposed scheme. A self-adaptive DE (SaDE) algorithm was proposed by Qin et al. [179] for adaptively controlling the scale factor (F) and crossover ratio (CR) in DE. SaDE employed four mutation strategy to generate donor vectors where stochastic universal selection technique is used to generate donor vector corresponding to target vector in the current population. The F is approximated using a normal distribution with mean 0.5 and standard deviation 0.3. On the other hand, CR is adapted by normal distribution $N(CRm, 0.1)$ with mean CR_m (which is initially set at 0.5) and standard deviation equal to 0.1. Zaharie [180] proposed a parameter adaptation for DE (ADE) based on the idea of controlling the population diversity and implemented a multi-population framework. Mallipeddi et al. [181] proposed an ensemble based mutation strategies and control parameters in the DE algorithm, called EPSDE. In the method, a pool of mutation strategies and values for each control parameter coexists throughout the evolution process and competes to produce offspring. Adaptive control DE was proposed in [182]. Gaussian distribution is used to adaptively control the parameters of the DE algorithm. Research in the recent past years has been evolved for PSP using the DE algorithm. However, parameter adaptation strategy of DE is missing while solving PSP problem.

In the chapter, an adaptive parameter control mechanism is proposed within the basic framework of the DE algorithm known as Lévy distributed DE (LdDE) [92] to control the scale factor and crossover ratio. Lévy distribution with variable location parameter has been invoked because it produces a significant amount of changes in the control parameters which ensures good exploration and exploitation in the search space to reach the global optimum solution. At first, experiments are carried out on well-known unimodal, basic and expanded multi-modal and hybrid composite benchmark functions with different dimensions. The proposed scheme (LdDE) is then applied to PSP problem. The real protein instances are varied from smaller length to higher length. Results are demonstrated the better performance of the LdDE in terms of accuracy and convergence speed compared to five parameter controlled variants DE algorithm.

The chapter is organized as follows. Section 5.2 explains the proposed LdDE algorithm. Experimental results demonstrating the performance of the LdDE in comparison with the conventional DE and several state-of-the-art adaptive DE variants over a fifteen benchmark functions are presented in Sect. 5.3. Section 5.4 provides an investigation of the proposed method over the real protein sequences. Section 5.5 summarizes the chapter.

5.2 Lévy Distributed DE (LdDE)

This section explains the proposed modification of the DE algorithm where the parameters such as F and CR are controlled adaptively in the execution phase by using Lévy distribution. The proposed LdDE algorithm employs the same crossover and selection technique as in the conventional DE algorithm which has been described earlier in the Sect. 1.7.3.1.

5.2.1 Lévy Distribution

The sum of two of normal random values is itself a normal random variable, which depicts that if X is normal, then for X_1 and X_2, independent copies of X and for any positive constants a and b, it is written in Eq. (5.1).

$$ax_1 + bx_2 \equiv cx + d, \tag{5.1}$$

where c and d are real positive values ($c, d \in \mathbb{R}$). The symbol $\equiv$ indicates equality in distribution, i.e. both expressions follow the same probability law. Physically, Eq. (5.1) represents the shape of X is preserved (up to scale and shift) under addition operation. A random variable X is stable if for X_1 and X_2, two independent copies of X, Eq. (5.1) is satisfied. Stable distributions are a rich class of probability distributions that allow skewness and heavy tails with many intriguing mathematical properties. The class was characterized by Paul Lévy [183] in his study of sums of independent identically distributed terms in the 1920s. In general, a Lévy Stable probability density function (PDF) [184, 185] is defined using Fourier transform of the characteristic function as:

$$\rho(t; \alpha, \beta, \mu, \sigma) = F(p(x; \alpha, \beta, \mu, \sigma))$$

$$= \int_{-\infty}^{\infty} e^{itx} p(x; \alpha, \beta, \mu, \sigma)\,dx$$

$$= exp[i\mu t - \sigma^{\alpha}|t|^{\alpha}(1 - i\beta\frac{t}{|t|}\omega(t, \alpha))], \tag{5.2}$$

where,

$$\omega(x, \alpha) = \begin{cases} \tan\frac{\Pi\alpha}{2} & \alpha \neq 1, \\ -\frac{2}{\Pi}\ln|x| & \alpha = 1. \end{cases} \tag{5.3}$$

Now $p(x; \alpha, \beta, \mu, \sigma)$ is calculated by Inverse Fourier transform of characteristics function $\rho(t; \alpha, \beta, \mu, \sigma)$, as follows:

$$p(x; \alpha, \beta, \mu, \sigma) = \frac{1}{2\pi} \int_{-\infty}^{\infty} \rho(t; \alpha, \beta, \mu, \sigma)\,dt. \tag{5.4}$$

In general, one can see that the characteristic function and the probability density function (pdf) are determined by the four real parameters: α, β, μ, and σ. The exponent $\alpha \in [0, 2]$ is the index of stability and the Lévy index, $\beta \in [0, 1]$ is the skewness parameter, μ is the shift parameter, and $\sigma > 0$ is a scale parameter. The indices α and β play a major role in the proposed approach. In particular, the pdf is expressed in terms of elementary functions as follows:

- Normal or Gaussian distribution, $\alpha = 2$,

$$f_N(x; \mu, \sigma) = \frac{1}{\sigma\sqrt{2\pi}} exp\left(-\frac{(x - \mu)^2}{2\sigma^2}\right). \tag{5.5}$$

- Cauchy distribution, $\alpha = 1$, $\beta = 0$,

$$f_C(x; \mu, \sigma) = \frac{\sigma}{\pi[\sigma^2 + (x - \mu)^2]}. \tag{5.6}$$

- Lévy–Smirnov distribution, $\alpha = 1/2$, $\beta = 1$,

$$f_L(x; \mu, \sigma) = \sqrt{\frac{\sigma}{2\pi}} \frac{1}{(x - \mu)^{\frac{3}{2}}} exp\left(-\frac{\sigma}{2(x - \mu)}\right). \qquad (5.7)$$

Both normal distributions and Cauchy distributions are symmetric, bell-shaped curves. The main qualitative distinction between them is that the Cauchy distribution [186] has much heavier tails. In contrast to the normal and Cauchy distributions, the Lévy distribution is highly skewed, with all of the probability concentrated on $x > 0$, and it has even heavier tails than the Cauchy.

5.2.2 The Lévy Distributed DE Algorithm

An interesting characteristic has been observed that for a small value of CR, a smaller number of components of an individual are changed in each generation and stepwise movement of vectors tends to be orthogonal to the current positions. On the other hand, high values of CR (near 1) keep most of the directions of the mutant vectors unaltered prohibiting the generation of orthogonal direction from the current location. Therefore, conventional DE performance is rotationally invariant only when CR equal to 1. A low CR value ensures search in each direction separately. In the study, a probability based adaptation scheme is employed to control the parameters of the DE algorithm. From the definition of heavy-tailed distribution, it has been observed that Lévy probability distribution function based on pseudo-random number generator generates a number that creates a considerable amount of diversity in the individual. A new heavy tail distribution is used for the scale factor and crossover ratio. The reason for choosing Lévy distribution is that heavy tail is able to generate a considerable amount of changes more frequently and it has a higher probability of performing a long jump that may escape from local optima or move away from a plateau that might provide a better solution for multi-modal optimization problems. In the proposed study, the initial range of scale factor, $F \in [0.5, 0.9]$ and crossover ratio, $CR \in [0.5, 1]$ are considered and parameters are updated time to time according to the Lévy distribution within the specified range. In the proposed algorithm, two constants $delF$ and $delCR$ are used for updating the shift parameters of the Lévy distribution to obtain updated control parameter values. The shift parameter of the Lévy distribution for two control parameters (F) and (CR) are $scaleF$ and $scaleCR$ which are initialized randomly within the range 0 to 1. The pseudo-code of the proposed scheme is described in Algorithm 8.

Algorithm 8 LdDE Algorithm

1: Initialize population of NP individuals randomly. Initialize $scaleF$ and $scaleCR$ randomly within the range 0 to 1
2: Evaluate fitness values of each individuals
3: **while** (Termination condition is not satisfied) **do**
4:    /* update scaling factor F/*
5:    **if** $delF < rand(0, 1)$ **then**
6:       $scaleF = scaleF + (Xmax - Xmin) \cdot rand(0, 1)$
7:       $F = Fmin + (Fmax - Fmin) \, Lévy \, (generation, 0.5, 1, 1, scaleF)$
8:    **else**
9:       $F = Fmin + (Fmax - Fmin) \, Lévy \, (generation, 0.5, 1, 1, scaleF)$
10:   **end if**
11:   **if** $F > Fmax$ **then**
12:      $F = Fmin + 0.4 \, rand(0, 1)$
13:   **end if**
14:   /* update crossover ratio CR/*
15:   **if** $delCR < rand(0, 1)$ **then**
16:      $scaleCR = scaleCR + (Xmax - Xmin) \, rand(0, 1)$
17:      $CR = CR + CRmax \, Lévy \, (generation, 0.5, 1, 1, scaleF)$
18:   **else**
19:      $CR = CR + (CRmax - CRmin) \, Lévy \, (generation, 0.5, 1, 1, scaleF)$
20:   **end if**
21:   **if** $CR > CRmax$ **then**
22:      $CR = CRmax \, rand(0, 1)$
23:   **end if**
24:   Generate donor vector $\mathbf{v}_i$ for each target vector $\mathbf{x}_i$ using *"DE/best/1/bin"* strategy and updated F value.
25:   Generate trial vector $\mathbf{u}_i$ using the binomial crossover and updated CR value
26:   Evaluate trial vector $\mathbf{u}_i$
27:   **if** $f(\mathbf{u}_i) \leq f(\mathbf{x}_i)$ **then**
28:      Replace $\mathbf{x}_i$ with $\mathbf{u}_i$
29:   **end if**
30: **end while**

5.3   Experimental Results

5.3.1   Benchmark Functions

A set of fifteen unconstrained well-known benchmark functions are taken for comparative study of the proposed method. Among the fifteen functions, f_1, f_2, f_3, f_4, and f_5 are unimodal functions and the remaining functions are multi-modal. The detail informations of the test functions such as name of the function, search space and optimum value are given in Table 5.1. The detailed description can be found in [187].

Table 5.1 The CEC 2005 real-parameter Benchmark Functions. Please refer to the details of the website http://www.ntu.edu.sg/home/EPNSugan/

Function No.	Name of the Test Function	Search Space	(f_{min})
f_1	Shifted Sphere Function	$[-100, 100]^D$	-450
f_2	Shifted Schwefels Problem 1.2	$[-100, 100]^D$	-450
f_3	Shifted Rotated High Conditional Elliptic Function	$[-100, 100]^D$	-450
f_4	Shifted Schwefels Problem 1.2 with Noise in fitness	$[-100, 100]^D$	-450
f_5	Schwefels Problem 2.6 with Global Optimum on Bounds	$[-100, 100]^D$	-310
f_6	Shifted Rosenbrocks Function	$[-100, 100]^D$	390
f_7	Shifted Rotated Griewanks Function Without Bounds	No Bounds	-180
f_8	Shifted Rotated Ackleys Function With Global optimum on bounds	$[-32, 32]^D$	-140
f_9	Shifted Rastrigins Function	$[-5, 5]^D$	-330
f_{10}	Shifted Rotated Rastrigins function	$[-5, 5]^D$	-330
f_{11}	Shifted Rotated Weierstrass Function	$[-0.5, 0.5]^D$	90
f_{12}	Schwefels Problem 2.13	$[-\pi, \pi]^D$	-460
f_{13}	Expanded Extended Griewanks plus Rosenbrocks Function	$[-3, 1]^D$	-130
f_{14}	Shifted Rotated Expanded Scaffers F6	$[-100, 100]^D$	-300
f_{15}	Hybrid Composite Function	$[-5, 5]^D$	120

5.3.2 Simulation Strategy

Performance of the proposed LdDE algorithm is compared with *DE/rand/1/bin* (RandDE), *DE/best/1/bin* (BestDE), *DE/Current to best/1/bin* (CurrBestDE), Adaptive Control DE (ACDE) [182] and SaDE [179] in 10 and 30 dimensions on each function. In order to make a fair comparison, we fixed the same seed for random number generation so that initial population is same for all the algorithms. In the experiment, the population size and the maximum number of function evolutions (FEs) are set as 50 and $10,000 \times D$ (D is the dimensions of the problem) for all the algorithms. In the analysis, we performed 50 independent runs for each functions. The parameter values of the RandDE, BestDE, CurrBestDE are same as defined in [188]. Default control parameter settings are used for ACDE [182] and SaDE [179]. In the proposed method, initially *delF* and *delCR* are set to 0.5. All algorithms are terminated either when function evaluations reached to defined FEs or achieved absolute error (e). The error is defined as $e = |f(\mathbf{x}) - f(\mathbf{x}^*)| \leqslant \theta$, where $f(\mathbf{x})$ is the current best solution and $f(\mathbf{x}^*)$ is the global optimum and θ is the error accuracy level. The error accuracy level, (θ) for all the algorithms is set to 10^{-6} for the functions f_1–f_5 and 10^{-2} for the remaining functions.

5.3.3   Results and Discussion

5.3.3.1   For 10-Dimension

Table 5.2 presents the *best*, *mean* and *standard deviation (std.)* of the error values over 50 runs for each benchmark functions with dimension 10 using six algorithms. The obtained best results of the six algorithms are shown in bold face.

From the results, it is observed that the LdDE algorithm performs better than all other algorithms for the functions f_1, f_2, f_3, f_4, f_5, f_6, f_7, f_8, f_{11}, f_{12}, f_{13}

Table 5.2 Results of the LdDE and other compared algorithms over fifteen test functions in 10-Dimension

F No.	Evolution Metrics	LdDE	RandDE	BestDE	CurrBestDE	ACDE	SaDE
f_1	Best	**3.82E-07**	5.68E-07	5.47E-07	5.64E-07	4.36E-07	3.88E-07
	Mean	**6.21E-07**	8.10E-07	8.17E-07	8.16E-07	7.77E-07	8.20E-07
	Std.	**1.65E-08**	9.88E-07	1.23E-07	1.48E-07	1.61E-07	1.82E-07
f_2	Best	**4.78E-07**	8.02E-04	5.37E-07	5.21E-07	5.16E-07	5.97E-07
	Mean	**7.12E-07**	3.36E-03	8.14E-07	8.43E-07	6.46E-06	8.80E-07
	Std.	**1.06E-07**	1.75E-03	1.39E-07	1.38E-07	2.81E-05	1.38E-07
f_3	Best	**9.72E-07**	2.71E+04	1.10E-01	6.33E-02	3.77E+02	1.90E+01
	Mean	**3.03E+01**	5.99E+04	5.31E+01	7.20E+01	1.85E+04	4.46E+03
	Std.	**1.30E+02**	2.05E+04	1.70E+02	2.62E+02	1.76E+04	6.58E+03
f_4	Best	**2.71E-07**	3.13E-02	5.07E-07	5.36E-07	4.57E-07	5.52E-07
	Mean	**8.13E-07**	2.51E-01	9.41E-07	8.94E-07	8.41E-07	1.23E-07
	Std.	**1.23E-07**	1.62E-01	4.65E-07	3.99E-07	1.50E-07	1.37E-07
f_5	Best	**5.78E-07**	1.20E-03	1.18E+03	3.47E+03	8.16E-07	7.23E-07
	Mean	**8.99E-07**	4.60E+01	5.03E+03	6.68E+03	1.11E-06	9.29E-07
	Std.	**1.06E-08**	1.09E+02	2.17E+03	2.00E+03	8.71E-07	6.67E-07
f_6	Best	**6.25E-03**	1.81E+00	2.60E+04	1.02E+02	2.04E-001	9.07E-03
	Mean	**4.86E-01**	2.41E+06	6.42E+06	6.37E+06	4.35E+00	5.01E-01
	Std.	**2.52E-01**	1.20E+07	1.67E+07	1.23E+07	1.96e+00	1.08E+00
f_7	Best	**5.98E-04**	6.38E-04	1.29E+00	7.22E-04	1.25E+03	1.27E+03
	Mean	**1.27E+03**	1.40E+03	1.57E+03	2.32E+03	1.32E+03	1.37E+03
	Std.	**3.88E-13**	5.80E-04	1.56E+02	1.74E+02	5.78E-10	5.82E-09
f_8	Best	**2.01E+01**	2.02E+01	2.03E+01	2.02E+01	2.03E+01	2.02E+01
	Mean	**2.02E+01**	2.05E+01	2.05E+01	2.05E+01	2.04E+01	2.05E+01
	Std.	**6.21E-02**	7.05E-02	1.17E-01	7.62E-02	7.04E-02	7.81E-02
f_9	Best	7.71E-03	9.99E-01	6.05E-03	4.56E+00	9.08E-03	**5.49E-03**
	Mean	3.73E+00	4.06E+00	5.19E+00	1.00E+01	6.71E+00	**8.02E-03**
	Std.	4.89E+00	1.72E+00	1.07E+00	4.39E+00	1.98E-01	**1.41E-02**

(continued)

Table 5.2 (continued)

F No.	Evolution Metrics	LdDE	RandDE	BestDE	CurrBestDE	ACDE	SaDE
f_{10}	Best	4.98E+00	4.29E+00	3.98E+00	9.07E+00	2.99E+00	**1.99E+00**
	Mean	1.28E+01	1.20E+01	1.89E+01	1.71E+01	8.84E+00	**5.73E+00**
	Std.	7.91E+00	4.90E+00	9.84E+00	4.48E+00	3.54E+00	**2.25E+00**
f_{11}	Best	**8.32E-03**	1.90E+00	9.85E-02	3.44E-01	5.07E-01	9.71E-02
	Mean	**1.69E+00**	8.40E+00	2.82E+00	4.01E+00	3.64E+00	3.00E+00
	Std.	**1.42E+00**	1.16E+01	2.60E+00	1.29E+01	9.93E+00	8.46E+00
f_{12}	Best	**5.89E-03**	1.37E+01	6.56E-03	1.78E+01	1.98E+01	9.57E-03
	Mean	**6.58E+00**	1.10E+03	3.03E+02	6.53E+02	4.86E+02	2.69E+02
	Std.	**7.61E+00**	1.07E+03	5.70E+02	7.72E+02	6.85E+02	6.19E+02
f_{13}	Best	**3.30E-03**	2.55E-01	7.08E-01	5.58E-01	8.77E-02	1.38E-01
	Mean	**2.23E+00**	8.57E-01	9.51E-01	5.47E-01	5.51E-01	7.94E+06
	Std.	**8.61E-02**	4.75E-01	6.08E-01	1.59E-01	1.38E-01	1.46E-01
f_{14}	Best	2.20E+00	1.74E+00	2.28E+00	1.76E+00	**1.33E+00**	2.23E+00
	Mean	**2.07E+00**	3.49E+00	3.08E+00	2.58E+00	2.80E+00	2.95E+00
	Std.	**2.58E-01**	4.84E-01	4.01E-01	3.72E-01	4.96E-01	3.14E-01
f_{15}	Best	1.35E+02	2.08E+00	7.17E-03	7.49E+00	**1.85E-03**	4.47E-03
	Mean	4.88E+02	2.72E+02	2.72E+02	2.72E+02	4.97E+01	**2.55E+01**
	Std.	1.86E+02	1.78E+02	1.68E+02	1.50E+02	4.29E+01	**3.22E+01**

and f_{14}. The *best* result for the function f_7 is achieved by LdDE while ACDE and SaDE produced the same result. Also, the ACDE algorithm achieved the *best* result than LdDE for the function f_{14}. The SaDE performed better than the LdDE for the functions f_9, f_{10} and f_{15} but ACDE dominates all other algorithms in terms of *best* result on the function f_{15}. The LdDE algorithm performs better compare to other algorithms for the functions f_7, f_8, f_{13} and f_{14} while accuracy level is poor with respect to mean error value. However, the LdDE algorithm has been achieved the global optimum with certain accuracy level on most of the test functions compare to all other algorithms. Comparing the results among the DE variants including the proposed LdDE, SaDE exhibits good search ability. The SaDE algorithm follows an adaptation of control parameters in mutation strategy by obtaining a balance performance between exploration and exploitation. The ACDE algorithm yields a comparatively better performance.

Table 5.3 presents results of unpaired t-tests between the best algorithm and the second best algorithm for each function in terms of two mean standard error difference, 95% confidence interval of this difference, the t-value, the two-tailed P value and the significance. For all the cases, sample size is 50 and degree of freedom is 98. It can be observed from Tables 5.2 and 5.3 that in most of the cases the proposed method meets or beats the nearest competitor in a statistically meaningful way. According to the results of t-tests, the LdDE is extremely statistically significant for

Table 5.3 Results of unpair t-test on the data of Table 5.2

F No.	Rank of Algorithm		Error	t	95% Conf. Interval	P-value	Significance
	1st	2nd					
f_1	LdDE	ACDE	0.00	4.832	(−2.21E-7, −9.10E-7)	<0.0001	Extremely Significant
f_2	LdDE	BestDE	0.00	3.01	(−1.70E-7, −3.40E-8)	0.0042	Very Significant
f_3	LdDE	BestDE	42.88	0.532	(−1.09E2, 6.34E1)	0.5971	Not Significant
f_4	LdDE	ACDE	0.00	0.707	(−1.05E-7, 5.00E-8)	0.4828	Not Significant
f_5	LdDE	SaDE	0.00	0.230	(−2.98E-7, 2.37E-7)	0.8189	Significant
f_6	LdDE	SaDE	0.22	0.069	(−4.59E-1, 4.29E-1)	0.9451	Not Significant
f_7	LdDE	ACDE	0.00	43.25E10	(−5.00E1, −5.00E1)	<0.0001	Extremely Significant
f_8	LdDE	ACDE	0.02	12.245	(−2.67E-1, −1.92E-1)	<0.0001	Extremely Significant
f_9	SaDE	LdDE	1.04	0.313	(−2.41,1.76)	0.7554	Not Significant
f_{10}	SaDE	LdDE	0.84	3.702	(−4.79,−1.41)	0.0006	Extremely Significant
f_{11}	LdDE	BestDE	0.59	1.901	(−2.31, 6.45E-2)	0.0633	Not Quite Significant
f_{12}	LdDE	SaDE	123.92	2.121	(−5.12E2, −1.37E1)	0.0391	Significant
f_{13}	LdDE	ACDE	0.03	5.828	(−2.55E-1, −1.24E-1)	<0.0001	Extremely Significant
f_{14}	LdDE	CurrBestDE	0.09	5.554	(−6.82E-1, −3.19E-1)	<0.0001	Extremely Significant
f_{15}	SaDE	ACDE	10.73	2.253	(−4.57,−2.60)	0.0288	Significant

the functions f_1, f_7, f_8, f_{13}, f_{14} and SaDE is extremely significant on the function f_{10} though LdDE takes the 2nd rank. Finally, the proposed method provides significant improvements in most of the test cases.

5.3.3.2   For 30-Dimension

The experiments conducted on fifteen test functions are repeated with considering dimension 30 for investigating the scalability of the proposed algorithm and the results are presented in Table 5.4.

Table 5.4 Results of the LdDE and other compared algorithms over fifteen test functions in 30-Dimension

F No.	Evolution Metrics	LdDE	RandDE	BestDE	CurrBestDE	ACDE	SaDE
f_1	Best	**7.48E-07**	9.77E+01	2.09E+04	2.60E+04	7.95E-07	7.85E-07
	Mean	**9.14E-07**	7.75E+02	4.14E+04	4.19E+04	2.43E-06	9.31E-07
	Std.	**5.08E-08**	5.46E+02	9.96E+03	8.88E+03	7.63E-08	7.56E-06
f_2	Best	**8.82E-07**	2.82E+03	3.20E+04	2.07E+04	1.26E-02	9.87E-07
	Mean	**9.69E-07**	8.87E+03	4.91E+04	3.04E+04	1.79E+00	9.55E-06
	Std.	**3.05E-08**	3.50E+03	1.43E+04	5.68E+03	5.93E+00	2.47E-05
f_3	Best	**2.58E+03**	2.56E+06	5.39E+07	1.23E+08	3.19E+05	2.24E+05
	Mean	**1.57E+05**	1.97E+07	4.79E+08	3.30E+08	1.29E+06	5.01E+05
	Std.	**2.06E+05**	1.19E+07	2.19E+08	1.62E+08	6.21E+05	3.14E+05
f_4	Best	**9.81E-07**	4.68E+03	3.52E+04	1.71E+04	2.04E+01	1.35E+00
	Mean	**6.35E-01**	9.79E+03	6.06E+04	2.81E+04	2.62E+02	9.99E+01
	Std.	**1.80E+00**	2.92E+03	1.63E+04	5.92E+03	3.17E+02	1.13E+02
f_5	Best	**1.03E+02**	3.06E+03	1.80E+04	2.15E+04	1.72E+03	2.34E+03
	Mean	**1.28E+03**	6.39E+03	2.72E+04	2.80E+04	2.39E+03	3.51E+03
	Std.	**3.54E+02**	1.59E+03	3.72E+03	2.56E+03	8.14E+02	8.22E+02
f_6	Best	**8.69E-03**	2.76E+06	9.30E-03	8.86E-03	2.52E+01	5.81E-01
	Mean	**1.44E-01**	1.71E+08	2.95E+01	2.89E+01	5.70E+01	4.45E+01
	Std.	**4.83E-01**	2.41E+08	2.80E+00	7.29E+00	3.29E+01	3.14E+01
f_7	Best	**4.69E+03**	4.86E+03	7.12E+03	8.91E+03	**4.9E+03**	**4.69E+03**
	Mean	**4.49E+03**	5.06E+03	9.17E+03	1.04E+04	4.69E+03	4.99E+03
	Std.	**2.91E-12**	1.27E+02	7.03E+02	5.54E+02	2.95E-12	2.99E-12
f_8	Best	2.09E+01	2.08E+01	2.05E+01	2.08E+01	2.08E+01	**2.03E+01**
	Mean	2.09E+01	2.99E+01	2.09E+01	2.17E+01	2.59E+01	**2.09E+01**
	Std.	**2.98E-02**	4.23E-02	1.27E-01	4.32E-02	1.46E-01	4.38E-02
f_9	Best	3.98E+00	9.40E+00	1.86E+02	1.12E+02	7.18E-03	**7.05E-03**
	Mean	4.82E+01	1.44E+02	2.55E+02	2.83E+02	**2.85E-01**	3.18E+00
	Std.	2.92E+01	4.91E+01	4.43E+01	1.98E+01	**5.35E-01**	2.58E+00
f_{10}	Best	3.58E+01	3.95E+01	2.53E+02	2.28E+02	**1.89E+01**	3.58E+01
	Mean	7.65E+01	6.29E+01	4.40E+02	3.06E+02	5.97E+01	**5.16E+01**
	Std.	3.85E+01	1.66E+01	8.64E+01	5.28E+01	1.38E+01	**1.32E+01**
f_{11}	Best	1.32E+01	8.39E+00	2.77E+01	1.71E+01	**8.09E+00**	1.31E+01
	Mean	2.41E+01	1.90E+01	3.36E+01	2.04E+01	**1.29E+01**	1.59E+01
	Std.	5.60E+00	2.99E+00	2.55E+00	3.43E+00	**1.66E+00**	2.65E+00
f_{12}	Best	**2.99E-01**	1.44E+04	4.77E+01	1.38E+01	3.34E+01	9.54E+01
	Mean	**2.05E+03**	4.66E+04	3.35E+03	3.24E+03	3.78E+03	3.68E+03
	Std.	**2.85E+03**	1.88E+04	4.31E+03	3.02E+03	2.98E+03	3.86E+03

(continued)

Table 5.4 (continued)

F No.	Evolution Metrics	LdDE	RandDE	BestDE	CurrBestDE	ACDE	SaDE
f_{13}	Best	1.37E+00	2.29E+00	9.88E+00	3.48E+00	**1.24E+00**	1.47E+00
	Mean	**3.08E+00**	1.77E+01	1.12E+01	1.07E+01	3.89E+00	4.62E+00
	Std.	**1.31E+00**	3.16E+01	7.48E+01	1.73E+00	2.06E+00	2.74E+00
f_{14}	Best	**1.09E+01**	1.23E+01	1.28E+01	1.13E+01	1.16E+01	1.18E+01
	Mean	**1.05E+01**	1.23E+01	1.33E+01	1.21E+01	1.33E+01	1.26E+01
	Std.	**1.86E-01**	5.26E-01	5.20E-01	2.70E-01	4.96E-01	2.69E-01
f_{15}	Best	**1.74E+01**	2.01E+02	6.18E+02	2.62E+01	3.18E+01	2.19E+02
	Mean	**3.37E+02**	4.03E+02	8.63E+02	3.41E+02	3.41E+02	3.77E+02
	Std.	**7.07E+01**	1.06E+02	1.15E+02	1.14E+02	8.54E+01	8.90E+01

From the results, it can be observed that LdDE surpasses other algorithms for the functions f_1, f_2, f_3, f_4, f_5, f_6, f_7, f_{12}, f_{13}, f_{14} and f_{15}. The SaDE performs better than other algorithms on functions f_8 and f_{10} as well as the ACDE achieves better results on functions f_9 and f_{11}. All 30-dimension functions become more difficult than the 10-dimension counterparts and the results are not as good as in 10-dimension cases with respect to solution accuracy, although we increased the maximum number of FEs from 100000 to 300000. The same *best* and *mean* the result is achieved by the algorithms LdDE, ACDE and SaDE for the function f_7. On the function f_{13}, the *best* error is achieved by ACDE but *mean* and *std.* are better in LdDE compared to all other algorithms. The search behavior of LdDE provides a balance between exploration and exploitation as observed by the results of the LdDE algorithm.

The unpaired t-tests between the best and the second best algorithm among six algorithms are presented in Table 5.5. Extreme statistical significant improvements has been observed for the functions f_3, f_4, f_5, f_6 and f_{14} by the LdDE algorithm. It can be observed in Table 5.5 that the ACDE significantly improves the results for the functions f_9 and f_{11}. The function f_8 significantly improves the SaDE algorithm. However, the LdDE algorithm performs significantly better compared to all other algorithms on fifteen test functions.

By analyzing the results of the LdDE algorithm over fifteen test functions in 10 and 30-dimension, one may conclude that the LdDE does not perform the best for the shifted rotated functions which are multi-modal. According to the 'no free lunch' theorem [100], any one algorithm cannot perform well over all the problems. Tuning of LdDE is needed to obtain better performance for the Shifted Rotated function. Therefore, we may not expect the best performance on all the problems as the proposed LdDE algorithm focuses on adaptively controlling the scale factor (F) and crossover ratio (CR), the parameters of DE.

Table 5.5 Results of unpair *t*-test for the data on Table 5.4

F No.	Rank of Algorithm		Error	*t*	95% Conf. Interval	P value	Significance
	1st	2nd					
f_1	LdDE	SaDE	0.00	1.13E-2	(−3.05E-6, 3.02E-6)	0.991	Not Significant
f_2	LdDE	SaDE	0.00	1.73	(−1.85E-5, 1.342E-6)	−8.58E-6	Quite Significant
f_3	LdDE	SaDE	7.52E4	4.57	(−4.95E5, −1.92E5)	0.001	Extremely Significant
f_4	LdDE	SaDE	22.62	4.38	(−1.44E2, −5.38E1)	0.001	Extremely Significant
f_5	LdDE	ACDE	179.18	1.24E1	(−2.58E3, −1.86E3)	0.001	Extremely significant
f_6	LdDE	CurrbestDE	1.46	1.96E1	(−3.17E1, −2.58E1)	0.001	Extremely Significant
f_7	LdDE	RandDE	25.56	1.42E1	(−4.15E2, −3.12E2)	0.001	Extremely significant
f_8	SaDE	BestDE	4.19	4.97	(1.24E1, 2.92E1)	0.001	Extremely significant
f_9	ACDE	SaDE	0.52	5.49	(−3.96, −1.83)	0.001	Extremely significant
f_{10}	SaDE	ACDE	9.68	8.34E-1	(−2.80E1, 1.18E1)	0.412	Not Significant
f_{11}	ACDE	SaDE	0.63	4.83	(−4.29, −1.76)	0.001	Extremely Significant
f_{12}	LdDE	CurrbestDE	831.49	1.42	(−2.85E3, 4.86E2)	0.161	Not Significant
f_{13}	LdDE	ACDE	0.49	1.66	(−1.79, 1.68E-1)	0.103	Not significant
f_{14}	LdDE	CurrbestDE	0.06	2.34E1	(−1.67, −1.40)	0.001	Extremely Significant
f_{15}	LdDE	CurrbestDE	26.76	1.23E-1	(−5.71E1, 5.05E1)	0.903	Not Significant

5.4   Application to Protein Structure Prediction

5.4.1   Simulation Strategy

The proposed LdDE algorithm and other algorithms (used for comparison in Sect. 5.3) are implemented in MatLab R2010b environment and executed on an Intel Core 2 Duo CPU with 8 GB RAM running at 2.53 GHz with Windows 7 home premium platform for protein structure prediction. In order to evaluate the performance

Table 5.6 Details of amino acid sequences used in the experiment

PS No.	PSL	PDB ID	Sequence
RS1	13	1BXP	MRYYESSLKSYPD
RS2	16	1BXL	GQVGRQLAIIGDDINR
RS3	18	2ZNF	VKCFNCGKEGHIARNCRA
RS4	21	1EDN	CSCSSLMDKECVYFCHLDIIW
RS5	25	2H3S	PVEDLIRFYNDLQQYLNVVTRHRYX
RS6	29	1ARE	RSFVCEVCTRAFARQEALKRHYRSHTNEK
RS7	35	2KGU	GYCAEKGIRCDDIHCCTGLKCKCNASGYNCVCRKK
RS8	38	1AGT	GVPINVSCTGSPQCIKPCKDQGMRFGKCMNRKCHCTPK
RS9	46	1CRN	TTCCPSIVARSNFNVCRLPGTPEAICATYTGCIIIPGATCP GDYAN
RS10	60	2KAP	KEACDWLRATGFPQYAQLYEDFLFPIDISLVKREHDFLDR DAIEALCRRL NTLNKCAVMK

of the proposed algorithm LdDE, we use ten real amino acid sequences as shown in Table 5.6. The real protein instances are taken from Protein Data Bank and experimented on 2D AB off-lattice model. The protein sequences, transformed into the AB sequences according to the following classification. The amino acids I, V, P, L, C, M, A and G are classified as hydrophobic ones (A) and amino acids D, E, H, F, K, N, Q, R, S, T, W, and Y are hydrophilic ones (B) [189]. The selected sequences have different lengths which enabled us to analyze and compare the proposed algorithm with different relevant algorithms.

In the experiment, candidate solutions are encoded in bond angles which are bounded within the range $[-180°, 180°]$. The dimension D of the solution varies from sequence to sequence based on the respective protein sequence length (PSL), which set as $(n - 2)$ where n is the length of the protein sequence. The stopping condition of each algorithm is the maximum number of function evaluations (FEs). The FEs is set to Max_FES $= 10,000 \times D$ defined in [130] when solving continuous optimization problem. In order to provide the fair comparison, 30 independent runs are performed for each protein sequence. All the algorithms are executed with the same initial population in each run. The parameter settings of the proposed algorithm and other algorithms are same as provided in Sect. 5.3.

5.4.2 Results and Discussion

Table 5.7 presents the *best*, *mean* and *standard deviation (std.)* of energy values obtained by the proposed LdDE, RandDE, BestDE, CurrBestDE, ACDE and SaDE algorithms over 30 independent runs for ten real protein sequences. The best results achieved by the algorithms are shown in bold face in Table 5.7. The convergence graphs of the compared algorithms for nine real protein instances are depicted in

Table 5.7 Best, Mean and Standard deviation values obtained by LdDE compared to other algorithms on RPs

PS No.	PSL	Evolution Metric	LdDE	RandDE	BestDE	CurrBestDE	ACDE	SaDE
RS1	13	Best	**−2.303**	−1.494	−0.3112	−2.180	−2.245	−2.246
		Mean	−1.956	−1.337	0.107	−1.503	−1.846	**−2.030**
		Std.	0.423	**0.116**	0.245	0.452	0.376	0.210
RS2	16	Best	**−7.595**	−7.018	−2.227	−4.356	−7.018	−7.127
		Mean	**−6.852**	−5.238	−1.284	−3.644	−6.644	−6.439
		Std.	0.619	1.424	0.578	0.546	**0.323**	0.607
RS3	18	Best	**−7.082**	−5.361	−1.211	−3.371	−6.877	−5.436
		Mean	**−6.956**	−4.111	0.798	−2.686	−6.276	−5.357
		Std.	0.918	0.856	1.521	0.663	0.881	**0.045**
RS4	21	Best	−7.431	−6.597	−2.187	−3.615	**−8.030**	−7.620
		Mean	**−6.869**	−4.392	−0.865	−3.128	−5.012	−6.008
		Std.	**0.385**	1.570	0.822	0.546	1.872	1.539
RS5	25	Best	**−8.857**	−5.203	1.044	−4.364	−6.801	−6.280
		Mean	**−6.901**	−3.833	1.629	−2.851	−6.497	−5.086
		Std.	1.641	1.151	0.621	0.871	**0.257**	1.068
RS6	29	Best	**−8.727**	−7.839	0.467	−2.403	−8.117	−8.559
		Mean	**−7.227**	−5.260	10.221	−1.973	−6.974	−7.019
		Std.	1.314	1.471	19.537	**0.296**	1.046	1.505
RS7	35	Best	−16.372	−10.365	1.279	−7.682	**−17.226**	−16.265
		Mean	**−15.156**	−8.637	24.771	−5.240	−14.758	−14.666
		Std.	**0.751**	1.572	39.920	1.778	1.836	1.476
RS8	38	Best	**−19.619**	−11.921	3.102	−7.220	−18.289	−16.957
		Mean	**−18.991**	−9.534	10.81E3	−5.231	−16.617	−15.403
		Std.	**0.746**	1.522	12.24E3	1.261	1.429	1.228
RS9	46	Best	−31.926	−21.611	4.681	−12.683	**−33.583**	−31.541
		Mean	**−29.127**	−19.625	10.89E3	−8.756	−26.358	−26.685
		Std.	2.459	**1.710**	20.28E3	2.575	5.236	2.959
RS10	60	Best	**−20.847**	−10.883	14.26E3	−5.775	−19.317	−20.327
		Mean	**−18.622**	−8.612	86.11E3	−3.736	−17.368	−17.281
		Std.	2.799	2.579	11.22E3	1.362	**1.214**	2.083

Fig. 5.1. In the plots, FEs represented in logarithmic scale while best objective value in normal visible scale.

It can be observed from Table 5.7 that LdDE achieves the top performance on the *best* results on seven sequences whereas ACDE produced *best* results on the sequence RS4, RS7, and RS9. In particular, the performance of the proposed method significantly outperforms compared to all other algorithms for the sequences RS3, RS5, and RS8 with respect to the *best* result. In addition, *mean* results are achieved

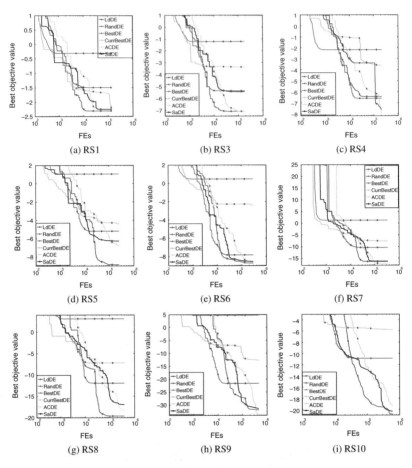

Fig. 5.1 Convergence characteristics of LdDE and other algorithms on nine real protein sequence

by LdDE is dominated the *mean* results of other algorithms in nine sequences except for the RS1. The SaDE algorithm achieved better results compared to the LdDE algorithm on the sequence RS1. Among ten real sequences, four sequences RS4, RS8, RS9 and RS10 provide significant results compared to others on the *mean* results. For remaining sequences, obtained results are very much close to the results of the SaDE or ACDE algorithm. It is surprising that the Lévy distributed DE algorithm obtained remarkable results in *mean* criteria on the higher length protein sequences. On the other hand, in terms of robustness, the LdDE algorithms is not a good one which is justified by the *std.* results of the sequences. However, the overall performance of the proposed methodology is better than the compared algorithms.

5.5 Summary

In the chapter, an adaptive mechanism is proposed to control the parameters of the DE algorithm for solving protein structure prediction problem as well as benchmark test functions with a different level of complexity. Lévy distribution is used to determine the scaling factor and crossover ratio in each generation during the execution process of the DE algorithm. The proposed algorithm is evaluated on both complex benchmark functions and real protein sequences with different lengths. The results are presented in the chapter demonstrate that the proposed scheme balanced between exploration and exploitation tread-off while solving multi-modal and PSP problems.

Chapter 6
Protein Structure Prediction Using Improved Variants of Metaheuristic Algorithms

Abstract This chapter introduces four schemes for protein structure prediction (PSP) based on 2D and 3D AB off-lattice model. The proposed methods are based on the modified versions of the classical PSO, BA, BBO and HS algorithms, providing an improved solution using different strategies. The strategies are developed in order to find global minimum energy value over the multi-modal landscape structure of the PSP problem. The performance of the proposed methods are extensively compared with the algorithms which are applied to the PSP problem over a test suite of several artificial and real-life protein instances.

6.1 Introduction

Protein structure prediction from amino acids sequence is a challenging task in computational biology. A stable three-dimensional structure of a protein always consumes minimum energy, represented by an energy function. The energy function can be formulated based on the physical model, here AB off-lattice model is employed. As discussed in Sect. 4.4, the nature of the energy landscape is highly rugged, contains many local optima i.e multi-modal in the continuous search space and have an influence on the performance of a metaheuristic search algorithm.

The particle swarm optimization (PSO), bees algorithm (BA), biogeography based optimization (BBO) and harmony search (HS) are metaheuristic techniques having stochastic behavior in the search process. Therefore, the algorithms are not free from premature convergence over a multi-modal landscape. The situation occurred due to loss of diversity in the population and solutions are being trapped in local optima. Various variants have been proposed [171, 190–192] in this context. Unfortunately, due to the characteristics of the PSP problem, solutions are mostly trapped in at least one local optima and fail to explore the search space. Therefore, in order to enhance

© Springer International Publishing AG 2018

N. D. Jana et al., *A Metaheuristic Approach to Protein Structure Prediction*, Emergence, Complexity and Computation 31,
https://doi.org/10.1007/978-3-319-74775-0_6

the exploration capability and to avoid being trapped into local optima, significant modifications are needed in the algorithmic framework which provides an improved performance for the PSP problem.

In this chapter, four variants of PSO, BA, BBO and HS are proposed for solving the PSP problem. (i) The PSO algorithm is modified with a local search, called PSOLS. In the PSOLS algorithm, Hill Climbing (HC) based local search is performed on each particle of the swarm to find the better solutions in the neighborhood search space. (ii) Next, an adaptive polynomial mutation based BA (APM-BA) has been proposed to improve the performance of BA. The mutation strategy is applied on best scout bees which do not improve the visited site in a predefined limit, known 'trial' counter of inefficient search. As a result, a high chance can be observed to jump out from the stacked visited site to unvisited site and made exploration on the search space. (iii) We enhanced population diversity of the BBO algorithm with chaotic mutation, named it BBO-CM algorithm in the book. Chaos system is used to generate a chaotic pseudo random number in mutation stage of BBO for exploring the solutions on the landscape of protein energy function. (iv) The HS algorithm is modified by using difference mean based perturbation mechanism, called as DMP-HS to each harmony to improved the performance of the basic HS for solving the PSP problem. In the proposed method, a new harmony vector is generated based on the best harmony and modified with normal distribution without using pitch adjustment rate (PAR) and the harmony memory is updated based on the greedy selection. In addition, a simple dimensional mean based perturbation scheme is integrated to all members of the harmony memory for better exploration and exploitation capabilities to ensure faster convergence and efficiency. All the proposed algorithms are investigated on artificial and real protein sequences with different length on the basis of 2D and 3D AB off-lattice model.

The rest of the chapter is organized as follows: Sect. 6.2 discusses the PSO algorithm with local search for PSP problem. Section 6.3 describes the adaptive polynomial based BA with experimental results. Section 6.4 explains the chaos induced mutation strategy for the BBO algorithm for solving the PSP problem. The HS algorithm with difference mean based perturbation procedure and experimental results for the PSP are presented in Sect. 6.5. Finally, a summary is drawn in Sect. 6.6.

6.2 Particle Swarm Optimization with Local Search

PSO was originally designed as a numerical optimization technique based on swarm intelligence and proved its robustness and efficiency to solving non-linear, real-valued function optimization problem. However, when dealing with complex problems, quality of the solution is affected by the number of generations increases and it suffers from premature convergences. The former situation occurs when all particles converge to a single point as the speed of the particles decreases with time. Thereafter, forcing them to converge to the global best point found so far which might not be global optima. PSO sometimes suffers from premature convergence due to many local minima in the search space. In general, convergence is a desirable property when swarms are allowed to search near the global minimum as time progresses. Unfortunately, in the context of many local minima, the convergence property may cause a swarm to be trapped in one of them and fail to explore more promising neighboring minima. To enhance the exploration capability and to avoid being trapped into local optima, a hybrid strategy is necessary to increase the diversity of the swarms in the search space. With this observation, a hybrid particle swarm optimization with local search algorithm has been proposed here. In the proposed approach, hill-climbing (HC) algorithm is used to execute a local search to find better solutions in the neighborhood search space produced by PSO at each generation.

6.2.1 Hill Climbing: A Local Search Algorithm

Hill Climbing (HC) is an optimization technique which belongs to the family of a local search algorithm. It is an iterative algorithm that starts with an arbitrary solution to a problem, then attempts to find a better solution by incrementally changing a single element of the solution with a neighborhood function (F_{ngh}). If the change produces a better solution, an incremental change is made to the new solution, repeating until no further improvements can be found or stop with a specific condition. The HC algorithm is used to local exploration for obtaining the local optimum (a solution that cannot be improved by considering a neighboring configuration) but it is not guaranteed to find the best possible solution (the global optimum) out of all possible solutions. HC local search algorithm is hybridized with particle swarm optimization providing more flexibility in the motion of particles. In the proposed algorithm, the neighborhood function (F_{ngh}) is designed to generate a new candidate solution, defined as follows:

$$New Solution(\mathbf{x}_{new}) = Current Solution(\mathbf{x}_c) + r_c(1 - 2\,rand()), \qquad (6.1)$$

where r_c represents the changing range of original particles and $rand(0, 1)$ is a random number between 0 to 1. Changing range is defined by Eq. (6.2).

$$r_c = a - (a - b)\frac{t}{t_{max}}, \qquad (6.2)$$

where a is the initial value of r_c and b is the value decreased by each generation t. t_{max} is the maximum number of generation. The pseudo-code of the Hill climbing algorithm is given in Algorithm 9.

Algorithm 9 Hill Climbing (HC) Algorithm

1: Initialize the current solution, $\mathbf{x}_c$ as the base point. solution size N
2: **for** $i = 1, 2, ..., N$ **do**
3:      Generate new solution $(\mathbf{x}_{new})$ using Eq. (6.1)
4:      Calculate $f(\mathbf{x}_{new})$ and $f(\mathbf{x}_c)$
5:      **if** $f(\mathbf{x}_{new}) \leq f(\mathbf{x}_c)$ **then**
6:          $\mathbf{x}_{new}$ is the best solution
7:          Set $\mathbf{x}_c = \mathbf{x}_{new}$
8:      **else**
9:          $\mathbf{x}_c$ as the best solution
10:     **end if**
11: **end for**

6.2.2   PSOLS Algorithm

In the proposed PSOLS algorithm, each particle is self-improved by applying local search algorithm before communicating with the population. Local search has been applied to all *pbest* (the best position found by each particle) of the swarm. In PSOLS algorithm, r_c is calculated using Eq. (6.2) which is linearly decreased as the generation increased. During the local searching, the particles are regarded to be matured. As the influence of the local searching, particles might be jumped out from a local optimum point and there is a greater chance to reach to the global optimum point. The proposed PSOLS algorithm is presented in Algorithm 10.

Algorithm 10 PSOLS Algorithm

1: Initialize swarm size (N), inertia weight (ω) and acceleration coefficients $(c_1$ and $c_2)$
2: Randomly initialize positions and velocities of N particles
3: Initially, local best position of each particle $(\mathbf{x}_{pbest})$ set as current positions of the particles
4: Evaluate fitness values of each particles
5: Calculate best particle in the swarm $(\mathbf{x}_{gbest})$
6: **while** Termination criteria is not satisfied **do**
7:      **for** $i = 1, 2, ..., N$ **do**
8:          Perform local search Algorithm 9 on particle $\mathbf{x}_i$ of the swarm
9:          Update personal best particle $\mathbf{x}_{ipbest}$ and global best particle $\mathbf{x}_{gbest}$ of the swarm
10:         Update velocity and position of the particles using Eqs. (1.15) and (1.14b)
11:         Evaluate fitness values of each particle
12:         Update $\mathbf{x}_{ipbest}$ and $\mathbf{x}_{gbest}$
13:     **end for**
14: **end while**

6.2.3 Experimental Results

Experiments are carried out on both artificial and real protein sequences to determine the protein structure prediction in 2D AB off-lattice model using the proposed PSOLS algorithm.

6.2.3.1 Artificial Protein Sequence

For the experiments, two kinds of artificial sequences are considered. Firstly, 20 artificial protein sequences of length 5 as in Stillinger [21] are considered, given in Table 6.1. Secondly, Fibonacci sequence is considered for obtaining an artificial protein sequence to predict protein structure in 2D AB off-lattice model [193]. The Fibonacci sequence is defined recursively as:

$$S_0 = A, \, S_1 = B, \, S_{i+1} = S_{i-1} * S_i$$

where $*$ is the concatenation operator. The first few sequences are $S_2 = AB$, $S_3 = BAB$, $S_4 = ABBAB$ and so on. Hydrophobic residue 'A' occurs in isolation along the chain, while hydrophilic residue 'B' occurs either isolated or in pairs and the molecules have a hierarchical string structure. Artificial protein sequence lengths are obtained through the Fibonacci numbers $n_{i+1} = n_{i-1} + n_i$. In the experiment, we consider artificial protein sequences, given in Table 6.2 with different lengths.

Table 6.1 Artificial protein sequences of length 5

Amino acid sequence			
AAAAA	AABAB	ABABB	BAABB
AAAAB	AABBA	ABBAB	BABAB
AAABA	AABBB	ABBBA	BABBB
AAABB	ABAAB	ABBBB	BBABB
AABAA	ABABA	BAAAB	BBBBB

Table 6.2 Artificial protein sequences with different lengths

AP	Length	Sequence
S_{13}	13	ABBABBABABBAB
S_{21}	21	BABABBABABBABBABABBAB
S_{34}	34	ABBABBABABBABBABABBABABBABABBABBAB
S_{55}	55	BABABBABABBABBABABBABABBABBABABBABBABABBABABBABBABABBABABB ABBABABBAB

Table 6.3 Real protein sequences

Real protein	Length	Sequence
1BXP	13	MRYYESSLKSYPD
1BXL	16	GQVGRQLAIIGDDINR
1EDP	17	CSCSSLMDKECVYFCHL
1EDN	21	CSCSSLMDKECVYFCHLDIIW
1AGT	38	GVPINVSCTGSPQCIKPCKDQGMRFGKCMNRKCHCTPK

6.2.3.2   Real Protein Sequences

In order to measure the effectiveness of the PSOLS algorithm in predicting the real protein structures, we select protein sequences from Protein Data Bank (PDB, http://www.rcsb.org/pdb/home/home.do). The PDB ID of these protein sequences are 1BXP, 1BXL, 1EDP, 1EDN and 1AGT respectively. The real protein (RP) sequence information are given in Table 6.3. In the experiment, the K-D method used in the literature [189] is adopted to distinguish the hydrophobic and hydrophilic residues among 20 amino acids in real protein sequences. The amino acids I, V, L, P, C, M, A and G are considered as hydrophobic ('A') residues and D, E, F, H, K, N, Q, R, S, T, W and Y are hydrophilic ('B') residues.

6.2.3.3   Parameter Settings and Initialization

The proposed PSOLS algorithm is compared with basic PSO [55] and PSO with a constriction factor (CPSO) [70], using artificial and real protein sequences with different lengths. The algorithms are implemented using MATLAB 7.6.0 (R2008a) applied on Intel (R) Core (TM) i7-2670QM CPU @ 2.20 GHz with 8 GB RAM on windows 7 Home Premium platform with same initial population but different number of generation based on the respective length of protein sequences. For the experiments, 5000, 12,000, 14,000, 16,000 and 20,000 generations of protein sequences of length 5, 13, 21, 34 and 55 are considered, respectively. Each algorithm runs 30 times and their *mean* and *standard deviation (std.)* are calculated. The population size for all the approaches are set as 50. The inertia weight (ω) and the acceleration coefficients (c_1 and c_2) for the basic PSO are set as $\omega = 0.732$ and $c_1 = c_2 = 1.49$ while for CPSO, $c_1 = c_2 = 2$. The parameters of the proposed PSOLS algorithm are set as $c_1 = c_2 = 2.05$ and inertia weight (ω) decreases linearly from 0.9 to 0.4 with increasing number of generations. In order to obtain less computational time spent by local search in the PSOLS algorithm, a small number (say 5) of neighbors of the current solution and r_c is initially set to 0.2 and its value decreases by 0.05 in consecutive generations.

Table 6.4 Results for artificial protein sequences with length 5

PS	PSO		CPSO		PSOLS	
	Mean	Std.	Mean	Std.	Mean	Std.
AAAAA	−3.017	0.089	−2.996	0.039	**−3.083**	0.000
AAAAB	−1.923	0.048	−1.903	0.012	**−1.931**	0.000
AAABA	−2.632	0.068	−2.583	0.069	**−2.675**	0.000
AAABB	−0.868	0.000	−0.880	0.004	**−0.888**	0.000
AABAA	−2.623	0.054	−2.576	0.037	**−2.655**	0.000
AABAB	−1.472	0.003	−1.452	0.010	**−1.474**	0.000
AABBA	−0.786	0.400	−0.901	0.035	**−0.962**	0.000
AABBB	0.040	0.000	0.041	0.001	**0.041**	0.000
ABAAB	−1.621	0.000	−1.692	0.016	**−1.721**	0.000
ABABA	−2.552	0.000	−2.476	0.036	**−2.562**	0.000
ABABB	−0.918	0.000	−0.951	0.004	**−0.958**	0.000
ABBAB	−0.118	0.137	−0.228	0.008	**−0.235**	0.049
ABBBA	−0.321	0.239	−0.448	0.016	**−0.475**	0.000
ABBBB	−0.044	0.023	−0.067	0.005	**−0.072**	0.012
BAAAB	−0.506	0.000	−0.520	0.003	**−0.526**	0.000
BAABB	0.086	0.000	0.091	0.001	**0.096**	0.000
BABAB	−0.832	0.000	−0.852	0.006	**−0.866**	0.000
BABBB	−0.490	0.000	−0.471	0.012	**−0.490**	0.000
BBABB	−0.312	0.000	−0.343	0.010	**−0.362**	0.000
BBBBB	−0.670	0.000	−0.657	0.010	**−0.680**	0.000

6.2.3.4 Results of Artificial Protein Sequences

Table 6.4 presents the results of *mean* and *standard deviation (Std.)* over 30 runs on the 20 different artificial protein (AS) sequences of length 5. From the results it can be observed that the proposed method performs better than basic PSO and CPSO. Moreover, the proposed algorithm is more robust than basic PSO and CPSO on some artificial sequences. From Table 6.5, it is clear that *mean* of minimum energy obtained by the PSOLS algorithms dominates the mean of the basic PSO and CPSO for all the artificial protein instances with lengths 13, 21, 34 and 55 over 30 runs. Better performance is obtained in case of the PSOLS algorithm compared to the other algorithms. For the robust solution, the PSOLS algorithm is placed in second position, compared to basic PSO and CPSO by considering *standard deviation (std.)* for the first three artificial protein sequences. On the other hand, in case of higher length of artificial sequence, the PSOLS algorithm is robust than other algorithms. The convergence characteristics for each algorithm on the artificial protein sequences with lengths 13, 21, 34 and 55 are shown in Fig. 6.1. The PSOLS algorithm exhibits faster convergence than the basic PSO and CPSO algorithm.

Table 6.5 Results for artificial protein sequences with different lengths

APs	PSO		CPSO		PSOLS	
	Mean	Std.	Mean	Std.	Mean	Std.
S_{13}	−1.805	0.581	−0.919	**0.274**	**−2.738**	0.416
S_{21}	−3.851	1.022	−0.660	**0.220**	**−4.968**	0.589
S_{34}	−5.670	1.450	1.188	**0.285**	**−6.051**	1.080
S_{55}	−8.299	1.865	5474	1630	**−8.328**	1.207

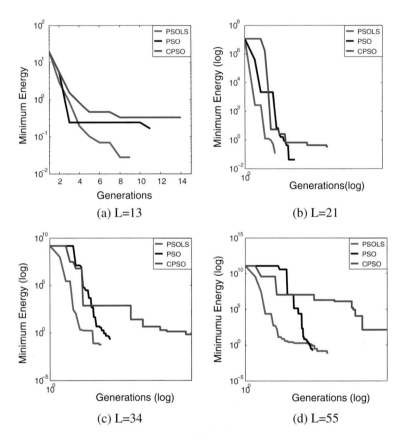

Fig. 6.1 Convergence characteristics of PSOLS, PSO and CPSO on artificial protein sequences with different lengths

Table 6.6 Results for real protein sequences with different lengths

RPs	PSO		CPSO		PSOLS	
	Mean	Std.	Mean	Std.	Mean	Std.
1BXP	−1.835	0.430	−0.883	0.220	**−2.423**	**0.0804**
1BXL	−6.468	1.300	−3.027	0.468	**−8.398**	**0.422**
1EDP	−3.563	1.448	−1.313	**0.481**	**−6.389**	0.5923
1EDN	−5.017	1.821	−0.805	**0.348**	**−7.095**	0.917
1AGT	−14.400	3.103	−0.049	**0.590**	**−16.965**	2.069

6.2.3.5 Results of Real Protein Sequences

The results of the experiments are reported in Table 6.6 representing the *mean* and *standard deviation (std.)* over 30 runs for the real protein instances such as 1BXP, 1BXL, 1EDP, 1EDN and 1AGT. It can be observed that lowest energy obtained by PSOLS algorithm is obviously better than the other algorithms like basic PSO and CPSO. Therefore, the proposed method provides better performance for solving real protein sequences. The lowest energy obtained by the PSOLS approach is highly robust than the basic PSO demonstrated by the standard deviation for all the real protein sequences. But robustness is slightly better than PSOLS in CPSO except for 1BXP real protein sequence. The convergence characteristics of PSOLS, basic PSO and CPSO on real protein sequences are depicted in Fig. 6.2, showing the faster convergence of PSOLS than the other two algorithms.

6.3 The Mutation Based Bees Algorithm

The Bees Algorithm (BA) (described in Sect. 1.7.4.2) was originally designed as a numerical optimization technique based on foraging behavior of honey bees and proved its robustness and efficiency to solving non-linear, real-valued function optimization problems. However, when dealing with multi-modal problems, quality of solution is affected as the number of generations increases and it suffers from premature convergence. The situation occurs when all the best visited sites converge in a small region of the search space, forcing them to converge to the global best point found so far which is not the actual global optima. BA sometimes suffers from premature convergence due to many local minima in the search space. In general, convergence is a desirable property that recruited bees of best visited sites and allowed to search near the global minimum as the time progresses. Unfortunately, in the context of many local minima, the scout bees of the best visited sites are trapped in one of the local minima and fail to explore more promising neighboring minima. To enhance the exploration capability and to avoid being trapped into local optima, a mutation strategy can be a suitable choice to increase the diversity of the best scout bees in

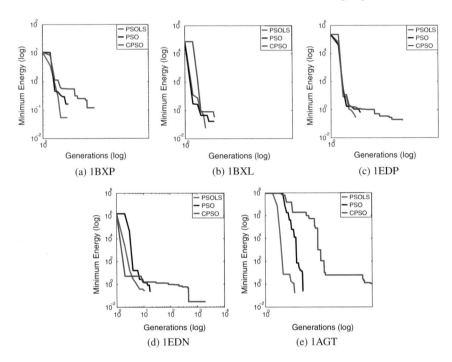

Fig. 6.2 Convergence characteristics of PSOLS, PSO and CPSO on real protein sequences

the search space. With this observation, an adaptive polynomial mutation (APM) is performed on the best bees of the BA, called APM-BA to prevent premature convergence when solving protein structure prediction problem. A neighborhood structure of each of the best selected site is determined using a local search processes in the proposed APM-BA algorithm. For example, the neighborhood $(x_{ij}(ngh))$ of the jth component of the ith best selected site is calculated as follows:

$$x_{ij}(ngh) = x_{ij} + rand(-2, 2). \qquad (6.3)$$

Consider a parameter *trial* representing the number of generations lead to inefficient search before better position is derived. If the ith best scout bee finds a better site, $trial(i)$ is set to zero; otherwise, it is incremented by one for the next generation. However, the searching competence of a best scout bee should not be evaluated by the quality of its current site i.e. the fitness value but by the efficiency of current search i.e. by the *trial* counter. Finally, to avoid premature convergence, APM is applied on the ith best scout bee when a predefined number of times the ith best scout bee cannot improve its current position.

6.3.1 Adaptive Polynomial Mutation Based BA

In adaptive polynomial mutation strategy [194], the jth dimension of the ith candidate solution $\mathbf{x}_i$ is mutated with the polynomial mutation is expressed using Eq. (6.4).

$$x_{ij}(t+1) = x_{ij}(t) + (x_j^{max} - x_j^{min})\delta_j, \qquad (6.4)$$

where t represents current generation number, x_j^{max} and x_j^{min} are the upper and lower bound of the jth component of the $\mathbf{x}_i$ while δ_j represents the polynomial function calculated using Eq. (6.5).

$$\delta_j = \begin{cases} (2r)^{\frac{1}{(\eta_m+1)}} - 1 & r < 0.5 \\ 1 - [2(1-r)]^{\frac{1}{(\eta_m+1)}} & r \geq 0.5 \end{cases}, \qquad (6.5)$$

where η_m is the polynomial distribution index and r represents uniformly distribute random number in $(0, 1)$. The probability of δ_j is calculated using Eq. (6.6).

$$P(\delta_j) = 0.5(\eta_m + 1)(1 - |\delta_j|)^{\eta_m}. \qquad (6.6)$$

By varying η_m, the perturbation is varied in the mutated solution. If the value of η_m is large, a small perturbation of a variable is achieved. To achieve gradually decreasing perturbation in the mutated solutions, the value of η_m is gradually increased. For each generation η_m is calculated using Eq. (6.7).

$$\eta_m = (80 + t). \qquad (6.7)$$

Finally, APM is applied on the ith best scout bees if the $trial(i) > D$ (dimension of the problem) for improving the diversity. The jth component of the ith mutated solution, mx_{ij} can be obtained through

$$mx_{ij} = x_{ij} + (x_{kj} - x_{ij})\delta_j. \qquad (6.8)$$

In Eq. (6.8), the ith best scout bee exchanges information with the kth one in its jth component where $k \neq i$. If $f(m\mathbf{x}_i) \leq f(\mathbf{x}_i)$, then ith best scout bee, $\mathbf{x}_i$ is replaced by mutated scout bee, $m\mathbf{x}_i$ and the $trial$ counter of the ith best scout bee set to 0.

The pseudo-code of APM-BA for the protein structure prediction is given in Algorithm 11. Since protein structure prediction is a minimization problem, fitness values are ranked in ascending order.

Algorithm 11 APM-BA Algorithm

1: Initialize the parameters of BA, Number of candidate solutions N, Dimension D and the inefficient *trial* counter
2: Initialize Population randomly and evaluate fitness values of each individual
3: Ranked fitness values $f(\mathbf{x})$ and the population
4: **for** $iter = 1$ to t_{max} **do**
5:   **for** $i = 1$ to E_s **do**
6:     **for** $j = 1$ to E_r **do**
7:       Generate neighborhoods according to Eq. (6.3)
8:     **end for**
9:     Select best neighborhood, $\mathbf{x}_{ngh}$
10:    **if** $f(\mathbf{x}_{ngh}) < f(\mathbf{x}_i)$ **then**
11:      Update $\mathbf{x}_i$ and $f(\mathbf{x}_i)$
12:        $trial(i) = 0$
13:    **else**
14:        $trial(i) = trial(i) + 1$
15:    **end if**
16:    **if** $trial(i) > D$ **then**
17:      Generate $m\mathbf{x}_i$ according to Eq. (6.8)
18:      **if** $f(m\mathbf{x}_i) < f(\mathbf{x}_i)$ **then**
19:        Update $\mathbf{x}_i$ and $f(\mathbf{x}_i)$
20:          $trial(i) = 0$
21:      **end if**
22:    **end if**
23:   **end for**
24:   **for** $i = 1$ to $(B_s - E_s)$ **do**
25:     **for** $j = 1$ to B_r **do**
26:       Repeat lines from 7 to 22
27:     **end for**
28:   **end for**
29:   **for** $i = 1$ to $(N - B_s)$ **do**
30:     Generate randomly and evaluate fitness $f(\mathbf{x}_i)$
31:   **end for**
32:   Ranked fitness values $f(\mathbf{x})$ and population
33: **end for**
34: Output the best solution

6.3.2 Experimental Results

In order to evaluate the performance of the APM-BA algorithm, the experiments are
performed on both artificial and real protein sequences described earlier in Tables 6.2
and 6.3 (in Sect. 6.2.3) for protein structure prediction using 2D AB off-lattice model.

6.3.2.1 Parameter Settings

The proposed APM-BA algorithm is compared with simple bees algorithm [171]
and the algorithms which are already used in protein structure prediction such as
CPSO [70], EPSO [71], IF-ABC [74] in 2D AB off-lattice model. All experiments

Table 6.7 APM-BA learning parameters

Parameters	Values
Number of scout bees (N)	50
Number of elite sites (E_s)	10
Number of best sites (B_s)	20
Number of recruited bees for elite sites (E_r)	10
Number of recruited bees for remaining best sites (B_r)	5

are implemented in MATLAB R2010a and executed on an Intel Core (TM) 17-2670 QMCPU running at 2.20 GHz with 8 GB of RAM with Windows XP. The independent experiments of each algorithm is repeated 30 times with the same initial population. The population size (N) for all approaches is fixed at 50 but the dimension D is different based on the respective length of protein sequences. The stopping criteria is same for all the algorithms based on number of generations (t_{max}). The number of generations is defined [130] by $t_{max} = (10,000 \times D)/N$. The parameters of the proposed algorithm are given in Table 6.7. The best of scout bees are going through adaptive polynomial mutation when *trial* counter of each of these scout bees are greater than D.

6.3.2.2 Results for Artificial Protein Sequences

Table 6.8 presents the results of *mean* and *standard deviation (std.)* over the 30 runs obtained by the APM-BA and CPSO, EPSO, IF-ABC and BA. From Table 6.8, it can be observed that the minimum energy obtained by the proposed algorithm dominates all other algorithms for every artificial protein sequences. Strong significant improvement has been observed in case of protein sequence of lengths 21 and 34 by the APM-BA. Three artificial protein sequences are placed in rank of 4th, 3rd and 2nd position with respect to standard deviation by the APM-BA which measures robustness of the algorithmic efficiency. Therefore, the proposed algorithm outperforms other algorithms with the increase of the length of artificial protein sequences. The convergence characteristics of each algorithm for the artificial protein sequences of lengths 13, 21 and 34 are shown in Fig. 6.3. The APM-BA exhibits faster convergence than other approaches.

6.3.2.3 Results for Real Protein Sequences

The results obtained by the APM-BA for real protein sequences are summarized in Table 6.9, along with the results of other algorithms used for comparison. It can be observed in Table 6.9 that the lowest energy obtained by the proposed algorithm is significantly better than the other algorithms like CPSO, EPSO, IF-ABC and

Table 6.8 Results for artificial protein sequences

APs	CPSO		EPSO		IF-ABC		BA		APM-BA	
	Mean	Std.	Mean	Std.	Mean	Std.	Mean	Std.	Mean	Std.
S_{13}	−1.653	0.518	−1.974	0.346	−1.846	**0.190**	0.020	0.228	**−2.596**	0.370
S_{21}	−3.352	0.741	−3.400	0.432	−3.062	**0.264**	1.491	0.850	**−4.964**	0.601
S_{34}	−5.284	1.302	−5.671	0.910	−4.586	**0.363**	2397.448	2845.384	**−6.694**	0.701

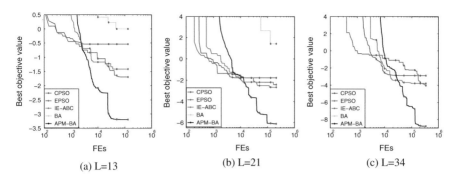

Fig. 6.3 Convergence characteristics of the algorithms on artificial protein sequences

Table 6.9 Results for real protein sequences

RPs	CPSO		EPSO		IF-ABC		BA		APM-BA	
	Mean	Std.	Mean	Std.	Mean	Std.	Mean	Std.	Mean	Std.
1BXP	−1.893	0.432	−2.110	0.250	−1.517	0.162	−0.118	0.349	**−2.335**	**0.085**
1BXL	−6.296	1.062	−6.255	1.161	−5.333	0.359	−0.696	1.087	**−8.002**	**0.592**
1EDP	−3.092	1.593	−3.773	1.438	−3.093	0.386	0.689	0.614	**−6.230**	**0.518**
1EDN	−4.232	1.325	−5.005	0.812	−4.232	0.482	2.135	1.058	**−7.275**	**1.078**

BA. Therefore, the APM-BA is superior in solving real protein sequences. Based on *standard deviation* results for all the real protein sequences compare to other algorithm, the proposed approach is highly robust. The convergence characteristics of the algorithms on real protein sequences are shown in Fig. 6.4, exhibits faster convergence rate in favor of the proposed APM-BA algorithm.

6.4 The Chaos-Based Biogeography Based Optimization Algorithm

The Biogeography Based Optimization (BBO) algorithm has already exhibited its robustness and efficiency to solving non-linear, real-valued function optimization problems. However, when dealing with the complex or multi-modal real-world problems, quality of the solutions is affected by the increase of a number of generations and it suffers from premature convergence. This behavior has been attributed to the loss of diversity in the population due to many local minima in the search space. Unfortunately, in the context of many local minima, individuals are trapped in one of the local minima and fail to explore more promising search region. In order to enhance the exploration capability and to avoid being trapped into local minima, we modify the mutation strategy of the BBO algorithm by increasing the population

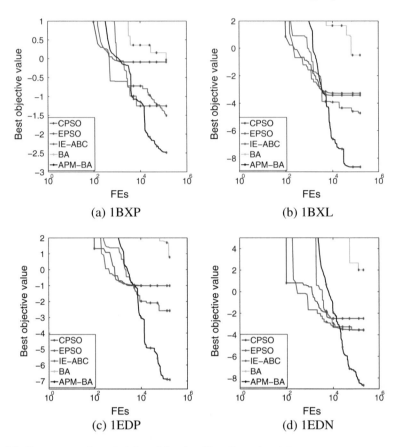

Fig. 6.4 Convergence characteristics of the algorithms for real protein sequences

diversity in the search space, based on the chaos theory. In the proposed mechanism, mutation probability (m_p) is selected adaptively within the user-defined range in each generation. Chaotic map generates the chaotic pseudo random sequence which is utilized for generating a variable of an individual of the population.

6.4.1 Chaotic Mutation in BBO

Chaos is defined as pseudo-random sequences generated by non-linear deterministic system [125, 126]. It has the properties of randomness, ergodicity, and regularity in order to traverse throughout the region of the search space. Chaotic systems have been used in evolutionary algorithms for its potential ability to escape from the local minima. Widely used one of the typical chaotic systems is the cubic map [126] which generates a chaotic sequence in [0, 1] defined in Eq. (6.9).

$$Z_{s+1} = 2.59 Z_s (1 - Z_s^2), \tag{6.9}$$

where s is the number of sequences to be generated.

In the proposed BBO-CM algorithm, mutation probability (m_p) is selected adaptively within a specified range $[a, b]$ for each dimension and calculated using Eq. (6.10).

$$m_p = a + (b - a) rand(0, 1). \tag{6.10}$$

In the mutation operation of the proposed BBO-CM algorithm, initially all the variables of ith solution $\mathbf{x}_i(x_{i,1}, x_{i,2}, ..., x_{i,D})$ are mapped to the variables $Z_i^0(z_{i,1}^0, z_{i,2}^0, ..., z_{i,D}^0)$ within the range $[0, 1]$ using the following formula, defined in Eq. (6.11).

$$z_{i,k}^0 = \frac{x_{i,k} - Min_i}{Max_i - Min_i}, \tag{6.11}$$

where $Min_i = min(\mathbf{x}_i)$, $Max_i = max(\mathbf{x}_i)$, $z_{i,k}^0$ is the initial chaotic random number of kth dimension of the ith solution. The initial chaotic numbers are used to generate chaotic pseudo random sequence in each dimension of the solution using Eq. (6.9). When kth dimension is to be mutated based on the mutation probability, the $z_{i,k}$, chaotic random number is used to replace the current kth dimension of the ith solution using Eq. (6.12).

$$x_{ik}^{new} = x_{ik}^{min} + (x_{ik}^{max} - x_{ik}^{min}) z_{ik}. \tag{6.12}$$

The proposed scheme explores the search space, prevents stagnant situations and improves in convergence speed by enhancing the population diversity. The pseudo-code of the chaotic mutation is presented in Algorithm 12. The pseudo-code of the proposed BBO-CM algorithm is given in Algorithm 13.

Algorithm 12 Chaotic Mutation Implementation

1: Initialize range of mutation probability (m_p) and (m_p) is calculated using Eq. (6.10)
2: **for** For each Habitat $i = 1, 2, ..., NP$ **do**
3:    Min_i and Max_i are determined
4:    Each variable of $\mathbf{H}_i$ are mapped to the variable of Z_i^0 using Eq. (6.11)
5:    **for** $j = 1, 2, ..., D$ **do**
6:      **if** rand$(0,1) < m_p$ **then**
7:        Chaotic pseudo random number sequence is generated using Eq. (6.9)
8:        Replace jth SIV of the $\mathbf{H}_i$ using Eq. (6.12) to make jth SIV of new $\mathbf{H}_i$
9:      **end if**
10:    **end for**
11:    Evaluate HSI for each habitat using protein energy function
12:    Replaced old $\mathbf{H}_i$ if new $\mathbf{H}_i$ is better
13: **end for**

Algorithm 13 BBO-CM algorithm

1: Initialize number of individuals (N), Dimensions (D)
2: Initialize Population randomly and evaluate HSI for each habitat.
3: **while** The stopping criterion is not satisfied **do**
4:     Short the population from best to worst
5:     Calculate the rate of λ_i and μ_i using Eq. (1.12)
6:     **for** each Habitat $i = 1, 2, ..., NP$ **do**
7:       Select habitat H_i with respect to rate λ_i
8:       **if** rand(0,1) $< \lambda_i$ **then**
9:         **for** each Habitat $j = 1, 2, ..., NP$ **do**
10:            Select habitat H_j with respect to μ_j
11:            Randomly select a SIV from H_j
12:            Exchange habitat information using Eq. (1.13)
13:         **end for**
14:       **end if**
15:     **end for**
16:     Calculated mutation probability for each habitat using Eq. (6.10).
17:     **for** $i = 1, 2, ..., NP$ **do**
18:       **for** $j = 1, 2, ..., D$ **do**
19:         **if** rand(0,1) $< m_p$ **then**
20:            Perform chaotic mutation using Algorithm 12
21:         **end if**
22:       **end for**
23:     **end for**
24:     Update the population
25:     Record the best habitat
26: **end while**

6.4.2 Experiments and Results

Experiments are carried out on same artificial and real protein sequences as described earlier in Tables 6.2 and 6.3 (in Sect. 6.2.3) to conform the efficiency of the proposed BBO-CM algorithm for protein structure prediction on 3D AB off-lattice model.

6.4.2.1 Experimental Settings

The proposed BBO-CM and other algorithms considered for comparisons are implemented in same system specification as in PSOLS and APM-BA. The population size (N) for all the approaches is fixed at 50 but the dimensions D is different and determined based on the respective length of the protein sequences which set to $(n - 2 + n - 3)$ (n is the length of the sequence). The number of maximum function evaluations, Max_FES $= 10,000 \times D$ is defined in [130] considered as stopping criteria for all algorithms. In BBO-CM, mutation probability (m_p) is adaptively selected within the range [0.01, 0.05]. In the experiment, parameters of BBO and BE-ABC are same as described in [52, 76].

6.4.2.2   Results and Discussion

The minimum energy values of the protein energy function, defined in Eq. (1.2) are obtained by using the BBO-CM, simple BBO and BE-ABC algorithms and presented in Table 6.10. The performance metrics, *best*, *mean* and *standard deviation (std.)* are considered over 30 independent runs. Best result obtained by the algorithms are in boldface in Table 6.10. From Table 6.10, it is cleared that the minimum energy value obtained by the BBO-CM are significantly better than those obtained by BBO, BE-ABC for the sequences with length 13 and 21. However, strong significant improvement has been observed in the case of the protein sequence with length 13. In case of the protein sequence with length 34, BE-ABC is better w.r.t *best* and *mean* value than BBO-CM and BBO algorithm. Although the best results obtained by BBO-CM is very near to the best results of BE-ABC, the BBO-CM algorithm is more robust compared to other algorithms in terms of standard deviation shown in Table 6.10. Therefore, the BBO-CM algorithm outperforms on APs with length 13 and 21; and comparable performance can be observed in the sequence of length 34. The convergence characteristics of each algorithm on APs with length 13, 21, and 34 are shown in Fig. 6.5. The BBO-CM exhibits faster convergence than the BBO and BE-ABC algorithms on each artificial protein sequences.

Table 6.11 represents the results of the algorithms over 30 independent runs in terms of *best*, *mean* and *std.* for real protein instances. It is observed that the minimum energy obtained by the BBO-CM algorithm outperforms other algorithms like BBO

Table 6.10 Results for artificial protein sequences

APs	BE-ABC			BBO			BBO-CM		
	Best	Mean	Std.	Best	Mean	Std.	Best	Mean	Std.
S_{13}	−3.685	−2.541	0.503	−0.614	−0.140	0.468	**−4.844**	**−2.874**	**0.383**
S_{21}	−6.096	−5.048	0.496	0.421	1.451	0.575	**−6.540**	**−5.569**	**0.319**
S_{34}	**−9.542**	**−7.028**	0.887	6.202	10.194	2.150	−9.167	−6.939	**0.697**

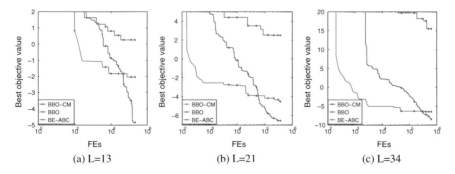

(a) L=13   (b) L=21   (c) L=34

Fig. 6.5 Convergence characteristics of the algorithms over artificial protein sequences based on 3D AB off-lattice model

Table 6.11 Results of real protein sequences with different algorithms based on 3D AB off-lattice model

RPs	BE-ABC			BBO			BBO-CM		
	Best	Mean	Std.	Best	Mean	Std.	Best	Mean	Std.
1BXP	−3.326	−2.732	0.197	−3.325	−2.637	0.392	**−3.680**	**−3.248**	**0.222**
1BXL	−11.118	−9.085	1.215	−11.574	−8.196	1.749	**−12.653**	**−10.642**	**1.058**
1EDP	−7.106	−5.676	1.027	0.119	1.141	0.455	**−7.509**	**−5.806**	**0.734**
1EDN	−9.695	−7.554	0.757	1.311	2.424	0.654	**−10.310**	**−7.824**	**1.106**
1AGT	−21.970	−18.352	2.065	−13.247	−8.139	2.123	**−22.095**	**−19.650**	**1.069**

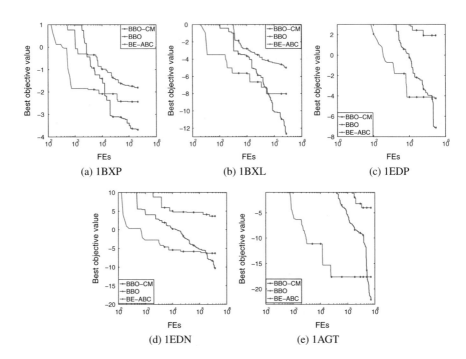

Fig. 6.6 Convergence characteristics of the algorithms for real protein sequences based on 3D AB off-lattice model

and BE-ABC for all the RPs. Therefore, the proposed approach indicates that it is an effective algorithm for minimizing the real protein structure in 3D AB off-lattice model. Moreover, robustness is highly significant of the BBO-CM compared to other algorithms w.r.t standard deviation. The convergence characteristics of the BBO-CM, BBO and BE-ABC on the RPs are given in Fig. 6.6 which reveals faster convergence of the BBO-CM algorithm compared to other algorithms.

6.5 The Difference Mean Based Harmony Search Algorithm

The Harmony Search (HS) algorithm has already proved its robustness and efficiency to solving numerical optimization problems. However, when dealing with complex or multi-modal numerical optimization problems like PSP problem, improvement in solutions decrease as the number of generations increases and it suffers from premature convergence. This situation occurred due to loss of diversity in the HM/population and solutions are trapped in local optima. Unfortunately, the characteristics of the PSP problem, solutions are mostly trapped in one of the local optima and fail to explore the search space. Therefore, in order to enhance the exploration capability and to avoid being trapped into local optima, an improved variant of HS algorithm has been proposed for PSP using 3D AB off-lattice model. In view of obtaining better results for the PSP problem, modifications and difference mean based perturbation (DMP) strategy is used to the basic HS algorithm and called it DMP-HS algorithm.

6.5.1 Difference Mean Based Perturbation (DMP) Method

The DMP strategy is used in particle swarm optimization (PSO) and shown its effective performance in global optimization [190]. The DMP is a learning strategy amounts to simply perturbing a newly generated population member with a scaled unit vector along any random direction. The unit vector is scaled by the difference mean formed by subtracting the dimensional mean of the individual to be perturbed from the current best individual of the population. Note that dimensional mean is a scalar quantity obtained by averaging the components of a vector individual from the population. After updating the HM, DMP is applied to all the harmony/individuals and pseudo-code of the DMP implementation is shown in Algorithm 14.

6.5.2 The DMP-HS Algorithm

The DMP-HS algorithm aims to overcome the limitations associated with the basic HS algorithm. In order to do so, some modifications have been made to the HS algorithm. In the proposed algorithm, pitch adjustment operation is not considered in such a way that the components chosen in new harmony vector with HMCR are directly modified. In the memory consideration process, jth component of the new harmony vector is selected from the jth component of the best harmony with HMCR. Then, the selected component of new harmony vector is modified as follows:

$$x_{ij}^{New} = x_{bestj} + bwN(0, 1),\qquad(6.13)$$

Algorithm 14 DMP implementation

1: Calculate the dimensional mean of the best individual ($\mathbf{x}_{best}$) with dimension, D:

$$best\_avg = \frac{1}{D} \sum_{j=1}^{D} \mathbf{x}_{best,j}$$

2: **for** each harmony $i = 1, 2, …, HMS$ **do**

3:    Calculate the dimension mean of ith individual, $\mathbf{x}_i$ of the population:

$$ind\_avg_i = \frac{1}{D} \sum_{j=1}^{D} x_{i,j}$$

4:    Calculate the difference mean of ith individual, $\mathbf{x}_i$:

$$diff\_mean_i = best\_avg - ind\_avg_i$$

5:    Generate a vector $\mathbf{p}$ with components chosen randomly unit normal distribution with zero mean and unit variance:

   $\mathbf{p} = \{p_1, p_2, …, p_D\}$

6:    Calculate $\hat{p}$: $\hat{p} = \frac{\mathbf{p}}{\|p\|}$

7:    Perturb amount of ith individual $\mathbf{x}_i$:

$$\mathbf{x}_i = \mathbf{x}_i + diff\_mean_i \times \hat{p}$$

8: **end for**

---

where bw is a distance bandwidth, the amount of changes or movement that may have occurred to the components of the new harmony vector. $N(0, 1)$ is a normal distribution with zero mean and unit variance. On the other hand, new vector components generated randomly with probability $(1 - HMCR)$. After improvisation of the new harmony vector, a greedy selection scheme is applied between $\mathbf{x}_i$ and $\mathbf{x}_i^{New}$ for updating the HM. DMP strategy is applied to all individuals after the HM updated to explore the search space for finding the global optimum solution. Finally, the pseudo code of the DMP-HS algorithm is presented in Algorithm 15.

6.5.3 Simulation Results

6.5.3.1 Protein Sequences

In order to find the better prediction results, the proposed DMP-HS algorithm is applied on real protein sequences (RPs) which are collected from Protein Data Bank (PDB, http://www.rcsb.org/pdb/home/home.do) and experimented on 3D AB off-lattice model. The detailed description of protein sequence length (PSL), PDB ID and sequence information of these real proteins (RS) are given in Table 6.12. The protein sequences, transformed into the AB sequences according to the classification rule in [189] as: the amino acids I, V, P, L, C, M, A and G are classified as hydrophobic ones ('A') and amino acids D, E, H, F, K, N, Q, R, S, T, W, and Y are hydrophilic ones ('B'). The selected sequences have different lengths which enabled us to analyze and compare the proposed algorithm with different relevant algorithms.

Algorithm 15 DMP-HS algorithm

1: Initialize the HM size (HMS), memory consideration rate (HMCR), Dimension (D)
2: Initialize the HM randomly and evaluate fitness values of each harmony.
3: **while** stopping criterion is not satisfied **do**
4:     Calculate best harmony $\mathbf{x}_{best}$ from the HM
5:     **for** each harmony $i = 1, 2, ..., HMS$ **do**
6:         **for** $j = 1, 2, ..., D$ **do**
7:             **if** $rand(0, 1) \leq HMCR$ **then**
8:                 component of new harmony vector, $x_{i,j}^{New}$ is generated using Eq. (6.13).
9:             **else**
10:                 $x_{i,j}^{New}$ is generated randomly
11:             **end if**
12:         **end for**
13:         **if** $f(\mathbf{x}_i^{New}) \leq f(\mathbf{x}_i)$ **then**
14:             $\mathbf{x}_i$ in the HM is replaced by $\mathbf{x}_i^{New}$
15:         **end if**
16:     **end for**
17:     **for** each harmony $i = 1, 2, ..., HMS$ **do**
18:         Perform DMP strategy using Algorithm 14
19:     **end for**
20:     Update the HM
21:     Record the best harmony so far
22: **end while**

Table 6.12 Details of amino acid sequences used in the experiment

PS No.	PSL	PDB ID	Sequence
RS1	13	1BXP	MRYYESSLKSYPD
RS2	16	1BXL	GQVGRQLAIIGDDINR
RS3	18	2ZNF	VKCFNCGKEGHIARNCRA
RS4	21	1EDN	CSCSSLMDK' ECVYFCHLDIIW
RS5	25	2H3S	PVEDLIRFYNDLQQYLNVVTRHRYX
RS6	29	1ARE	RSFVCEVCTRAFARQEALKRHYRSHTNEK
RS7	35	2KGU	GYCAEKGIRCDDIHCCTGLKCKCNASGYNCVCRKK
RS8	38	1AGT	GVPINVSCTGSPQCIKPCKDQGMRFGKCMNRKCHCTPK
RS9	46	1CRN	TTCCPSIVARSNFNVCRLPGTPEAICATYTGCIIIPGATCP GDYAN
RS10	60	2KAP	KEACDWLRATGFPQYAQLYEDFLFPIDISLVKREHDFLDR DAIEALCRRLNTLNKCAVMK

6.5.3.2   Parameter Settings

In the experimental study, all individuals are encoded in angles within the range $[-180°, 180°]$. The dimension, D of an individual varies from sequence to sequence based on the respective PSL which set as $(n - 2 + n - 3)$ (n is the length of the sequence). All the experiments are executed on the basis of stopping criterion which is the maximum number of function evaluations (FEs). The FEs is set to $\texttt{Max\_FES} = 10,000 \times D$ defined in [130] which is same for other algorithms used for comparisons. The DMP-HS algorithm is compared with CPSO, BE-ABC and the basic HS algorithm. The parameters for CPSO, BE-ABC and the basic HS algorithms are same as recommended in [76, 78, 192]. The parameters for the DMP-HS algorithm are set as follows: $HMS = 50$, $HMCR = 0.9$ and $bw = 0.2$. Each algorithm is run 30 times independently for each RPs with the same initial population.

6.5.3.3 Results and Discussion

Table 6.13 shows the *best, mean* and *standard deviation (std.)* of energy values achieved by CPSO, BE-ABC, HS and DMP-HS algorithm on the RPs with 30 independent runs. Wilcoxon's rank sum test [172] is conducted at the 5% significance level in order to judge whether the *mean* result obtained with the best performing algorithm differs from the *mean* result of the other compared algorithms in a statistically significant way. '+' marks in Table 6.13 indicate that DMP-HS statistically outperformed in comparison with other algorithms based on *mean* result. All marks are labeled with parenthesis in Table 6.13. The best results are shown in bold faced. The comparisons in Table 6.13 shows that the DMP-HS algorithm achieved the highest accuracy performance in terms of *best* and *mean* results on the RPs. In particular, the performance of the DMP-HS significantly outperforms compared to all other algorithms for higher length protein sequences (RS7, RS8, RS9 and RS10) in terms of *best* and *mean* results. In addition, results obtained by the DMP-HS algorithm outperforms on the remaining RPs. BE-ABC and HS algorithms achieved approximately same on the *mean* result for the sequence RS2, RS3, RS4, RS5, RS6 and RS7 whereas BE-ABC outperforms HS on RS8, RS9 and RS10. The merits of CPSO are worst than the others algorithms on all the sequences. The strong significant improvement has been observed on the DMP-HS algorithm than others as the protein sequence length increases. Moreover, obtained results are very robust based on standard deviation in case of DMP-HS algorithm except for RS5. Therefore, the proposed approach provides better performance and it is an effective algorithm for determining the structure of real proteins using 3D AB off-lattice model. Best convergence characteristics are obtained by the DMP-HS algorithm and other algorithms on RPs are shown in Fig. 6.7. The graph conclusively establishes the faster convergence of the DMP-HS algorithm over other algorithms. It is evident from the figures that only the proposed method has been able to obtain high-quality solutions.

Table 6.13 Comparisons of best, mean and standard deviation results for RPs

PS No.	PSL	Evolution Metric	CPSO	BE-ABC	HS	DMP-HS
RS1	13	Best	−0.802	−2.893	−4.150	**−4.498**
		Mean	−0.022 (+)	−2.160 (+)	−3.640 (+)	**−4.193**
		Std.	0.590	0.734	0.424	**0.306**
RS2	16	Best	−2.845	−9.949	−13.886	**−15.200**
		Mean	−0.932 (+)	−9.084 (+)	−9.534 (+)	**−14.880**
		Std.	1.007	0.695	2.415	**0.373**
RS3	18	Best	1.377	−8.103	−10.607	**−15.056**
		Mean	2.418 (+)	−5.815 (+)	−7.455 (+)	**−14.108**
		Std.	0.681	1.648	1.813	**1.096**
RS4	21	Best	3.397	−9.914	−10.660	**−17.721**
		Mean	4.622 (+)	−7.689 (+)	−8.005 (+)	**−16.519**
		Std.	0.900	0.866	1.518	**0.760**
RS5	25	Best	5.941	−7.670	−10.857	**−15.240**
		Mean	6.413 (+)	−6.803 (+)	−8.205 (+)	**−13.905**
		Std.	0.358	**0.834**	1.815	1.058
RS6	29	Best	5.766	−10.258	−12.277	**−17.416**
		Mean	7.120 (+)	−8.126 (+)	−9.827 (+)	**−15.965**
		Std.	1.004	1.002	1.290	**1.058**
RS7	35	Best	6.830	−20.891	−19.614	**−40.696**
		Mean	8.739 (+)	−18.612 (+)	−16.709 (+)	**−37.945**
		Std.	1.116	1.508	1.369	**0.901**
RS8	38	Best	6.775	−22.442	−18.616	**−40.300**
		Mean	9.086 (+)	−20.614 (+)	−16.277 (+)	**−39.637**
		Std.	1.207	1.410	1.641	**0.426**
RS9	46	Best	8.601	−39.998	−30.876	**−61.426**
		Mean	11.085 (+)	−38.008 (+)	−23.538 (+)	**−59.142**
		Std.	2.113	1.391	5.450	**1.083**
RS10	60	Best	20.730	−27.140	−13.142	**−44.972**
		Mean	23.675 (+)	−23.107 (+)	−8.991 (+)	**−42.634**
		Std.	1.572	1.967	2.678	**0.772**

6.6 Summary

In this chapter, particle swarm optimization with local search (PSOLS), adaptive polynomial mutation based bees algorithm (APM-BA), biogeography-based optimization with chaotic mutation (BBO-CM) and harmony search with difference mean perturbation (DMP-HS) are proposed for solving protein structure prediction problem more efficiently compare to the existing methods. The characteristics of

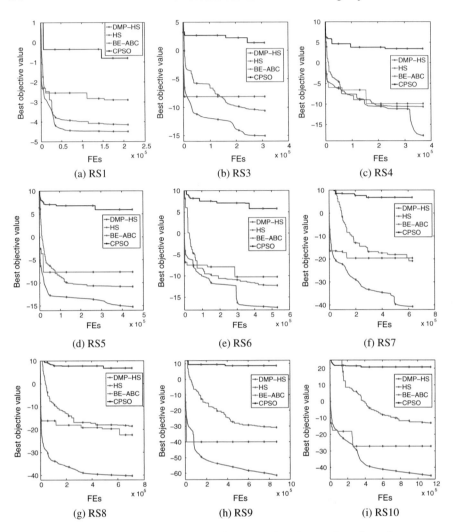

Fig. 6.7 Convergence curves of best objective value for the algorithms on the real protein sequences

the PSP problem, solutions are mostly trapped in one of the local optima and fail to explore the search space. In the PSOLS algorithm, a local search algorithm is applied on *pbest* of all the particles to preventing particles fall into local optima and made exploration on their search space. The proposed, PSOLS is investigated on artificial and real protein sequences using the 2D AB off-lattice model to evaluate the performance. In APM-BA, adaptive polynomial mutation is applied to the best scout bees based on their inefficiency during the search processes. The proposed strategy is able to prevent stuck at local optima and made exploration on the search space. Experimental results confirmed that APM-BA is efficient for protein structure prediction problem with respect to artificial and real protein sequences. The BBO-CM algorithm is configured for protein structure optimization based on 3D AB off-lattice model. Chaotic mutation in the BBO algorithm is able to jump out from a stagnant position in local optima and made exploration on their search space resulting in a diversity of the population and avoid premature convergence of the algorithm. The results presented in the chapter revealed the better performance of the proposed BBO-CM with respect to artificial and real protein sequences. Finally, in DMP-HS, all the harmony are perturbed through difference mean of the current harmony and best harmony to improve the exploration capability for preventing premature convergence and ability to jumping out from the local optima. The proposed algorithm is evaluated on the real protein sequences with varying sequence length using 3D AB off-lattice model. The analysis and experiments show that the DMP-HS algorithm is significantly effective for solving PSP problem.

Chapter 7
Hybrid Metaheuristic Approach for Protein Structure Prediction

Abstract Hybridization is an integrated framework that combines the merits of algorithms to improve the performance of an optimizer. In this chapter, the synergism of the improved version of particle swarm optimization (PSO) and differential evolution (DE) algorithms are invoked to construct a hybrid algorithm. The proposed method is executed in an interleaved fashion for balancing exploration and exploitation dilemma in the evolution process. The results are tested on ten real protein instances, taken from the protein data bank. The effectiveness of the proposed algorithm is evaluated through qualitative and quantitative comparisons with other hybridization of PSO and DE; and comprehensive learning PSO algorithms.

7.1 Introduction

Stochastic optimization algorithms have potential to solve optimization problems in various fields of engineering and science [195, 196]. However, increasing non-linearity, multi-modality, discontinuity and even dynamics make the problems more complex and intractable [197]. Unfortunately, traditional computing methods are inefficient to determine global optimum solution on the rough non-linear surface. Heuristic algorithms have been used for determining a global solution for such type of problems. A heuristic algorithm is knowledge dependent, so finding a unique heuristic optimization algorithm is difficult that produces an optimum solution for all problems. Hybridization is an integrated framework where merits of algorithms are utilized to improve the performance of the optimizers [198].

In the evolution process, the success of an algorithm depends on two important issues [199]: exploration (population diversity) and exploitation (selection pressure). These two factors are inversely related and it is important to mention that if the selection pressure increases, the search focuses only on the good individuals (in terms of fitness) resulting loss of diversity that ultimately leads to a premature convergence at a suboptimal solution. On the other hand, a lower value of selection pressure may

© Springer International Publishing AG 2018

N. D. Jana et al., *A Metaheuristic Approach to Protein Structure Prediction*, Emergence, Complexity and Computation 31, https://doi.org/10.1007/978-3-319-74775-0_7

not examine the search space properly resulting in stagnation. Therefore, it is challenging to balance the two factors for designing hybrid algorithm in order to obtain better performance. Particle swarm optimization (PSO) and differential evolution (DE) are two different types of population-based stochastic search algorithms each having its own limitations [51, 200]. In PSO, the major problem is slow convergence rate mainly in the latter period of generation due to lack of diversity in the search space. On the other hand, DE is good at exploring the search space and locating the region of global optimum, but it is slow at the exploitation of the solution [201]. Hybridization is one of the most efficient strategies where merits of PSO and DE algorithms are utilized to improve the performance of the optimizers. However, the existing PSO-DE based hybridization algorithm [202–204] is designed where modification is incorporated into the main operation, neglecting the problem of the exploration-exploitation dilemma in finding a solution to the multi-modal problems. In the proposed integrated framework, the balance between exploration and exploitation are considered in order to explore the search space efficiently for achieving the global optimum solution.

In this chapter, an integrated framework of hybridization has been proposed, called as hybrid PSO-DE (HPSODE) algorithm where interleaving between improved versions of PSO and DE are performed for solving multi-modal problems like protein structure prediction problem. In the HPSODE algorithm, the improved version of PSO is designed by incorporating adaptive polynomial mutation (APM) to the global best particle of PSO for increasing the diversity in the swarm. Similarly, the DE algorithm is improved by adopting trigonometric mutation to generate the donor vectors to achieve higher selection pressure. The proposed algorithm starts with improved PSO and switched to improved DE based on fitness value because PSO has higher ability during the initial period of generations, whereas DE performs better in the later period of searching to find more accurate solutions [205]. The HPSODE algorithm is tested on ten real protein instances. Based on the experimental results, it has been observed that HPSODE improves performance better in terms of efficiency, accuracy, and robustness.

The rest of the chapter is organized in the following way. Section 7.2 presents the proposed hybridization framework of the PSO and DE algorithms. Experimental setup and the results of the protein structure prediction are presented and analyzed in Sect. 7.3. Finally, the chapter is summarized in Sect. 7.4.

7.2 The Hybridization Framework

In HPSODE algorithm, an effective adaptive polynomial mutation (APM) strategy has been applied to PSO that prevents solutions being trapped at local optima. On the other hand, the trigonometry mutation operator is applied to DE for enhancing the performance of the algorithm. To achieve global optimum, an improved PSO and improved DE are used alternatively to develop HPSODE algorithm based on the fitness value obtained in successive generations.

7.2.1 Improved PSO (IPSO)

The basic PSO algorithm has been improved by applying APM [206], first introduced in GA based on polynomial probability distribution, formulated as follows:

$$x_{ij}(t+1) = x_{ij}(t) + (x_j^{max} - x_j^{min})\delta_j. \tag{7.1}$$

Where the jth component of an ith candidate solution $\mathbf{x}_i$ is mutated using polynomial mutation and t represents current generation number. x_j^{max} and x_j^{min} represent the upper and lower bound of the jth component of the ith solution $\mathbf{x}_i$ while the parameter δ_j can be calculated as follows:

$$\delta_j = \begin{cases} (2r)^{\frac{1}{(\eta_m+1)}} - 1 & r < 0.5 \\ 1 - [2(1-r)]^{\frac{1}{(\eta_m+1)}} & r \geq 0.5 \end{cases}, \tag{7.2}$$

where η_m is the polynomial distribution index and r represents uniformly distributed random number in $[0, 1]$. The probability of the parameter δ_j is defined as:

$$P(\delta_j) = 0.5(\eta_m + 1)(1 - |\delta_j|)^{\eta_m}. \tag{7.3}$$

Perturbation varies in the mutated solution by varying η_m value. If the value of η_m is large, a small perturbation is achieved while to achieve slow perturbation in decreasing order, the value of η_m is increased gradually in the mutated solutions. In [206], $\eta_m = (80 + t)$ (t is the current generation) is considered at which adaptation can be achieved, known as adaptive polynomial mutation.

To improve the performance of PSO, each component of the global best solution is mutated using APM as follows:

$$mgbest_j(t) = gbest_j(t) + (x_j^{max} - x_j^{min})\delta_j. \tag{7.4}$$

Here x_j^{max} is the upper bound and x_j^{min} is the lower bound of the jth component of the **gbest** of the population.

A selection operation is performed to determine whether the mutated global best (**mgbest**) or global best (**gbest**) survives to the next generation. The selection operation is performed using Eq. (7.5).

$$\mathbf{gbest}(t+1) = \begin{cases} \mathbf{mgbest}(t) & \text{if } f(\mathbf{mgbest}(t)) \leq f(\mathbf{gbest}(t)) \\ \mathbf{gbest}(t) & \text{otherwise} \end{cases}, \tag{7.5}$$

where $f(.)$ is the objective function to be minimized.

7.2.2 Improved DE (IDE)

Usually, in the DE algorithm, a mutation operator is performed on the basis of three individuals in the current population. In the mutation strategy, trigonometry mutation operator [207] is used to generate donor vector in DE. The three mutually exclusive individuals, $\mathbf{x}_{r_1}(t)$, $\mathbf{x}_{r_2}(t)$ and $\mathbf{x}_{r_3}(t)$ are randomly chosen to generate ith individual $\mathbf{x}_i(t)$, where $r_1, r_2, r_3 \in [1, 2, \ldots, NP]$ and $r_1 \neq r_2 \neq r_3 \neq i$. When a trigonometric mutation operation is performed, a donor to be perturbed is taken to the center point of the triangle. The perturbation imposed to the donor is then made up with a sum of three weighted vector differentials and defined in Eq. (7.6).

$$
\mathbf{v}_i(t+1) = \frac{\mathbf{x}_{r_1}(t) + \mathbf{x}_{r_2}(t) + \mathbf{x}_{r_3}(t)}{3} + (p_2 - p_1)(\mathbf{x}_{r_1}(t) - \mathbf{x}_{r_2}(t)) + \\
(p_3 - p_2)(\mathbf{x}_{r_2}(t) - \mathbf{x}_{r_3}(t)) + (p_1 - p_3)(\mathbf{x}_{r_3}(t) - \mathbf{x}_{r_1}(t)),
\tag{7.6}
$$

where $p = |f(\mathbf{x}_{r_1}(t))| + |f(\mathbf{x}_{r_2}(t))| + |f(\mathbf{x}_{r_3}(t))|$, $p_1 = \frac{|f(\mathbf{x}_{r_1}(t))|}{p}$, $p_2 = \frac{|f(\mathbf{x}_{r_2}(t))|}{p}$ and $p_3 = \frac{|f(\mathbf{x}_{r_3}(t))|}{p}$. In the proposed IDE, mutation operation is performed using Eq. (7.6) with a mutation probability τ_μ or using Eq. (7.7) with probability $(1 - \tau_\mu)$ to generate the donor vectors in the DE algorithm.

$$
\mathbf{v}_i(t) = \mathbf{x}_{r_1}(t) + F(\mathbf{x}_{r_2}(t) - \mathbf{x}_{r_3}(t)).
\tag{7.7}
$$

The mutation probability τ_μ is defined by the user. The flowchart of the improved PSO (IPSO) and improved DE (IDE) is given in Fig. 7.1.

7.2.3 The HPSODE Algorithm

In HPSODE algorithm, an improved PSO (IPSO) is executed at the beginning of each generation and the global best solution is mutated with the adaptive polynomial mutation to produce a better solution. However, after successive generations, if there is no change in the global best solution (i.e. global best solution gets stagnant at a point in the search space) then the process is switched to improved DE (IDE) to find the global best solution. Similarly, if the global best solution obtained by the improved DE is stacked over successive generations then the process is switched to IPSO. This process is repeated for the maximum number of generations in order to escape from local minimum by exploring search space at each generation. Say, $f^*(t) = f^*(t+1) = f^*(t+2) = \cdots = f^*(t+N)$, so execution is shifted to either IDE or IPSO based on the f^* (fitness) of the global best solution for N number of generations, up to which stagnation can be tolerated. In HPSODE algorithm when

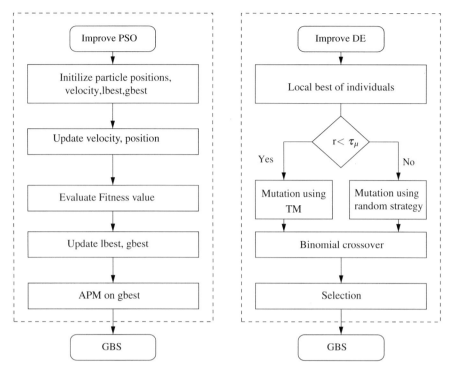

Fig. 7.1 Flow diagram of the improved PSO and improved DE algorithm. Here, lbest: local best of particle, gbest: global best in the swarm, r: random number, τ_μ: mutation probability, TM: trigonometry mutation, APM: adaptive polynomial mutation and GBS: global best solution

the positions of the particles are updated by the IDE, velocity of the particles are also updated. Otherwise, the velocity of the particles become irrelevant and sometimes misleading while switching to IPSO. The particle velocity is updated through the selection processes of the IDE using following Eq. (7.8).

$$\mathbf{v}_i(t+1) = \mathbf{v}_i(t) + (\mathbf{u}_i(t) - \mathbf{x}_i(t))rand(0, 1), \tag{7.8}$$

where $\mathbf{v}_i(t)$, $\mathbf{u}_i(t)$ and $\mathbf{x}_i(t)$ are the velocity of the ith particle, trial vector and target vector, respectively of the ith vector at generation t. $rand(0, 1)$ is the uniformly distributed random number in [0, 1]. The pseudo-code of the HPSODE algorithm is given in Algorithm 16.

Algorithm 16 HPSODE Algorithm

1: Initialize the particles position and velocity, C (Counter for same global best solution) = 0, P
   and Q are fixed integers
2: **while** (Stopping condition not satisfied or maximum no. of generation (t) not reached) **do**
3:   **if** ($t == 1$) **then**
4:     IPSO;
5:   **end if**
6:   **if** ($GBS(G - 1) == GBS(G)$) **then**
7:     $C = C + 1$;
8:   **else**
9:     IPSO;
10:     $C = 1$;
11:   **end if**
12:   **if** ($P \leq C \leq Q$) **then**
13:     IDE;
14:     Particles velocity updated during DE execution using Eq. (7.8)
15:   **else**
16:     IPSO;
17:     **if** ($C = P + Q$) **then**
18:       $C = P$;
19:     **end if**
20:   **end if**
21: **end while**

7.3 Experiments for PSP Problem

7.3.1 Simulation Strategy

7.3.1.1 Protein Sequences

In order to evaluate the efficiency of the proposed HPSODE technique in PSP problem solving, we use ten real protein sequences with different length in the experiment. The detail information about the protein instances such as PDB ID, sequence length and amino acid sequence are given in Table 7.1. The protein sequences, transformed into the AB sequences according to the following classification. The amino acids I, V, P, L, C, M, A and G are classified as hydrophobic ones ('A') and amino acids D, E, H, F, K, N, Q, R, S, T, W and Y are hydrophilic ones ('B') [189]. The selected sequences have different lengths which enabled us to analyze and compare the HPSODE algorithm with other hybrid algorithms based on 2D AB off-lattice model.

7.3.1.2 Parameter Settings and Initialization

To validate the effectiveness of the proposed algorithm, we compare the HPSODE algorithm with DEPSO-ZH [208], DEPSO [202], HPSO [203], DEPSO-CX [209] and CLPSO [210]. All experiments are carried out on a system with a Core (TM)

Table 7.1 Details information about amino acid sequences used in the experiment

PS No.	PSL	PDB ID	Sequence
RS1	13	1BXP	MRYYESSLKSYPD
RS2	16	1BXL	GQVGRQLAIIGDDINR
RS3	18	2ZNF	VKCFNCGKEGHIARNCRA
RS4	21	1EDN	CSCSSLMDKECVYFCHLDIIW
RS5	25	2H3S	PVEDLIRFYNDLQQYLNVVTRHRYX
RS6	29	1ARE	RSFVCEVCTRAFARQEALKRHYRSHTNEK
RS7	35	2KGU	GYCAEKGIRCDDIHCCTGLKCKCNASGYNCVCRKK
RS8	38	1AGT	GVPINVSCTGSPQCIKPCKDQGMRFGKCMNRKCHCTPK
RS9	46	1CRN	TTCCPSIVARSNFNVCRLPGTPEAICATYTGCIIIPGATCPGDYAN
RS10	60	2KAP	KEACDWLRATGFPQYAQLYEDFLFPIDISLVKREHDFLDRDAIEAL
			CRRLNTLNKCAVMK

17-2670 QMCPU running at 2.20 GHz with 8 GB of RAM. The operating system is MS Windows XP and implemented using MATLAB 7.6.0 (R2008a). Each algorithm runs 30 times independently and the same initial population is considered for each run. The population is initialized randomly within the range $[-180°, 180°]$. The stopping condition of each algorithm is fixed as the maximum number of function evaluations (*FEs*). The *FEs* is set to `Max_FES` $= 10,000 \times D$, defined in [130] when solving continuous optimization problem. Population size for all algorithms are set to $NP = 50$. The inertia weight (ω) and acceleration coefficients (c_1 and c_2) for the DEPSO, DEPSO-CX and HPSO are set as $\omega = 0.729$ and $c_1 = c_2 = 1.49$. For DEPSO, scale Factor (F) $= 0.2$ and Crossover ratio (CR) $= 0.5$. CLPSO set the default parameters as in [210]. The parameters of HPSODE algorithm is set as follows: The acceleration coefficients (c_1 and c_2), crossover ratio (CR) and mutation probability (τ_μ) are set as 2, 0.1 and in the range $[0.03, 0.05]$, respectively. On the other hand, inertia weight ω and scaling factor F are in decreasing nature as generation increases and calculated using Eq. (7.9).

$$Value = m - (m - n)\frac{t}{t_{max}}, \tag{7.9}$$

where m and n are constant while t and t_{max} represent the current generation and maximum generations respectively. For inertia weight (ω), $m = 0.9$ and $n = 0.5$. In case of F, $m = 1.5$ and $n = 0.3$. These parameters are chosen based on the literature [54]. Two integer P and Q in the decision processes of HPSODE are set as 50 and 100, respectively. The summary of the parameter settings for each of the algorithms are given in Table 7.2.

Table 7.2 The parameter settings of the algorithms

Algorithms	Parameters							
	c_1	c_2	ω	F	CR	τ_μ	P	Q
HPSODE	2.05	2.05	m = 0.9, n = 0.5	m = 1.5, n = 0.3	0.1	[0.03, 0.05]	50	100
DEPSO [202]	1.49	1.49	0.729	0.2	0.5	–	–	–
DEPSO-ZH [208]	2	2	0.4	–	0.001	–	–	–
DEPSO-CX [209]	2.05	2.05	0.729	–	0.1	–	50	–
HPSO [203]	1.49	1.49	0.729	0.5	0.2	–	–	–

7.3.2 Results and Discussion

Table 7.3 presents the *best*, *mean* and *standard deviation (std.)* of minimum energy values obtained by the HPSODE, DEPSO, DEPSO-ZH, DEPSO-CX, HPSO and CLPSO algorithms over 30 independent runs on real protein sequences using 2D AB off-lattice model. Minimum on the *best*, *mean* and *std.* results of the algorithm for the protein instances are in bold face shown in Table 7.3. Convergence characteristics of HPSODE along with other hybrid variants are shown for nine real protein sequences in Fig. 7.2.

An outstanding performance has been achieved by the HPSODE algorithm. The *mean* results are obtained by HPSODE dominates all other algorithms for each protein instances irrespective of sequence lengths. However, *mean* results of the DEPSO algorithm on sequence RS5 and RS6 are close to *mean* results of the HPSODE algorithm. It can be observed from Table 7.3 on the *best* results that the proposed HPSODE method significantly outperforms on the protein sequences, except the two sequences RS5 and RS9. It is noticed that CLPSO does not perform well any of the sequences, rather DEPSO performs better compared to DEPSO-ZH, DEPSO-CX, HPSO and CLPSO in terms of *best* and *mean* on the smaller as well as higher lengths of the protein sequences. The HPSODE algorithm is less robust compared to other algorithms judged by the obtained *std.* values. The overall performance of the HPSODE becomes effective for determining the structure of real protein sequences using 2D AB off-lattice model.

The plots in Fig. 7.2 correspond to the best objective value obtained in 30 runs of each algorithm against the function evaluations (FEs). In the plots, FEs are represented in logarithmic scale. It can be observed from Table 7.3 that *best* value achieved by HPSO is unexpected and far away from the *best* results of the other algorithms. Convergence plots provided in Fig. 7.2e–i also indicate the same fact.

Table 7.3 Results of the algorithms on real protein sequences with different lengths

PS no.	PSL	Evolution metric	HPOSDE	DEPSO	DEPSO-ZH	DEPSO-CX	HPSO	CLPSO
RS1	13	Best	**−2.429**	−2.196	−2.303	−2.191	−0.223	−1.403
		Mean	**−2.367**	−1.719	−2.203	−1.590	0.008	−0.556
		Std.	**0.087**	0.387	0.170	0.383	0.134	0.767
RS2	16	Best	**−8.341**	−7.127	−6.905	−6.620	−3.591	−4.983
		Mean	**−7.338**	−6.121	−5.872	−5.991	−1.812	−3.005
		Std.	1.039	0.591	1.012	**0.531**	1.045	1.417
RS3	18	Best	**−7.133**	−5.614	−6.640	−5.793	0.047	−2.405
		Mean	**−6.064**	−5.243	−5.641	−5.032	0.455	−1.189
		Std.	1.043	0.385	0.795	0.676	**0.373**	1.079
RS4	21	Best	**−8.250**	−4.459	−7.729	−5.751	1.456	−2.041
		Mean	**−6.145**	−3.455	−5.143	−4.237	1.644	−1.496
		Std.	1.459	0.703	1.770	0.850	**0.219**	0.359
RS5	25	Best	−7.007	**−7.562**	−5.971	−4.825	2.337	−2.233
		Mean	**−6.271**	−5.676	−5.119	−4.231	4.553	−0.358
		Std.	0.611	1.332	0.796	**0.393**	2.407	1.212
RS6	29	Best	**−8.854**	−8.588	−8.568	−5.346	34.006	−3.182
		Mean	**−7.148**	−6.847	−6.609	−4.716	28.876	43.741E2
		Std.	1.259	1.417	1.165	**0.535**	5.605	97.108E2
RS7	35	Best	**−16.793**	−16.311	−13.489	−10.621	56.058	−2.019
		Mean	**−15.034**	−14.003	−13.124	−9.465	97.970E1	30.551
		Std.	1.305	1.953	**0.454**	1.382	86.814E1	70.650
RS8	38	Best	**−21.015**	−18.531	−17.754	−11.217	86.388	−3.698
		Mean	**−18.281**	−16.514	−14.121	−8.897	64.894E1	40.584
		Std.	2.137	2.347	2.626	**1.355**	81.795	77.021
RS9	46	Best	−30.840	**−33.790**	−32.772	−19.989	84.602E3	0.430
		Mean	**−28.540**	−26.692	73.788E1	−16.720	15.277E4	37.116E3
		Std.	**1.803**	5.262	17.058E2	2.856	56.909E3	72.685E3
RS10	60	Best	**−21.468**	−20.610	−18.249	−9.801	49.637E4	1.144
		Mean	**−18.940**	−17.806	−14.120	−8.585	59.957 E4	63.748E4
		Std.	2.883	3.214	2.421	**0.741**	12.144E4	13.537E5

7.4 Summary

The chapter presents an integrated framework between improved version of PSO and DE, named as HPSODE for solving protein structure prediction problem. The proposed approach improves the searching abilities to find the global optimal solution with balancing between exploration and exploitation in the search space. Falling of particles into local optima have been prevented and driven away particles more effi-

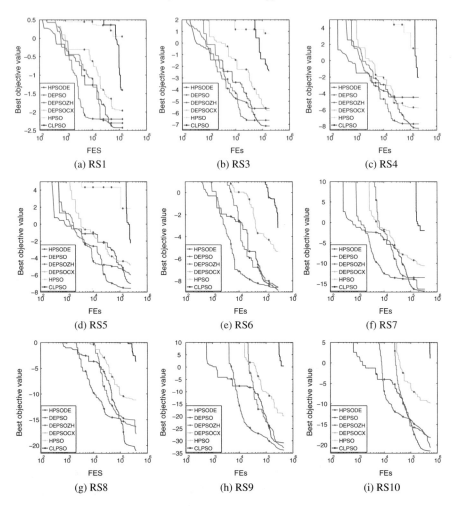

Fig. 7.2 Convergence characteristics of comparable algorithms over nine real protein sequences based on 2D AB off-lattice model

ciently by incorporating adaptive polynomial mutation on the global best solution in PSO. The performance of DE increases using the trigonometric mutation operator to create donor vector with prior probability. The proposed algorithm is experimented on ten real protein sequences with different length. The performances are evaluated based on the solution quality, convergence speed, and scalability analysis. The experimental results demonstrated the HPSODE algorithm outperforms for each protein instances and improves the performance in a significant way.

Chapter 8
Conclusions and Future Research

Abstract This chapter concludes the observations while solving protein structure prediction problem using a metaheuristic algorithm based on fitness landscape analysis. The achievements and also the shortcomings of the proposed methods are discussed in details. At the end future research ideas related to the work is also presented.

8.1 Discussions

Protein structure prediction from amino acids sequence is considered as one of the most crucial tasks in bioinformatics. The structure of a protein determines essential functions which play important role in medicine, drug design, disease prediction and much more. The *Anfinsens thermodynamic hypothesis* states that stable three-dimensional structure consumes a global minimum free energy surface of a protein. So, the structure prediction of a protein from its amino acids sequence is transformed into a global optimization problem. Metaheuristic based computational approach for protein structure prediction is a matter of interest among the researchers in recent decades due to the complexity of the problem. The major observations that can be drawn from each chapter are listed below.

- In order to gain an insight into the fitness landscape structure of the continuous search space, Chap. 3 of this book proposes a chaos induced random walk algorithm. The proposed random walk algorithm in continuous search space is used as the basis of techniques for characterizing the features of an optimization problem such as ruggedness, smoothness, modality, neutrality and deception. The distribution of chaotic random walk sample points over the search space is more uniform than the distribution of the simple random walk samples and hence provides a better coverage area of the entire search space. In addition, entropy and fitness distance correlation measures are used for investigating the ruggedness and deception of all the benchmark functions [130] test suite and some real protein instances with different length, both are defined in continuous search space.

© Springer International Publishing AG 2018
N. D. Jana et al., *A Metaheuristic Approach to Protein Structure Prediction*, Emergence, Complexity and Computation 31,
https://doi.org/10.1007/978-3-319-74775-0_8

- The key contribution of this book is presented in Chap. 4. Fitness landscape analysis is performed on 65 protein sequences of artificial and real protein instances to determine the characteristics of the problem or its structural features based on which the most appropriate algorithm is recommended for solving the protein structure prediction problem. The quasi-random sampling technique and city block distance is used to generate the landscape-path for the proteins instances on both 2D and 3D AB off-lattice model. Fitness landscape measures demonstrate that the PSP problem has highly rugged landscape structure containing many local optima and needle-like funnels with no global structure that characterizes the problem complexity. In addition, the problem complexity increases with the increasing sequence length irrespective of the type of the protein instances. Findings exhibit that the PSO algorithm outperforms on most of the protein instances compared to the other algorithms in 2D AB off-lattice model. The ABC algorithm significantly outperforms on all the protein instances compared to other algorithms in 3D AB off-lattice model.
- An adaptive parameter control metaheuristic algorithm is proposed in Chap. 5 for solving multi-modal problems. Lévy distribution is used to determine the scaling factor and crossover ratio of the DE algorithm in each generation during the execution process. Results are computed on standard benchmark functions and real protein instances with different length. The competency of the proposed method is evaluated and compared with five widely used variants of DE. The competency of the LdDE is judged using best and mean value of the objective function, standard deviation, student t-test and convergence speed.
- The four strategies PSOLS, AMP-BA, BBO-CM and DMP-HS are proposed for handling multi-modality in protein structure prediction presented in Chap. 6 of this book. The particle swarm optimization with local search (PSOLS), adaptive polynomial mutation based bees algorithm (APM-BA), biogeography base optimization with chaotic mutation (BBO-CM) and difference mean based harmony search algorithms have effectively enhanced the diversity in the population and avoid being trapped into local optima. The developed variants of the metaheuristic, algorithms were tested on extensive artificial and real protein sequences with 2D and 3D AB off-lattice model. The performance of the methods is measured in terms of accuracy, robustness and convergence speed via best value, mean value, standard deviation of the protein energy function and the convergence graphs. The new variants were shown to perform statistically better than several other metaheuristic algorithms [74, 76, 78] already used in protein structure prediction.
- In Chap. 7, the PSO and DE algorithms are combined to form a hybrid metaheuristic framework for solving the PSP problem. Adaptive polynomial mutation and trigonometry mutation strategies were utilized in PSO and DE that incorporate an explorative and exploitative behavior to balancing their search effects. The hybrid HPSODE algorithm is compared with the four developed hybrid variants of PSO and DE [202, 203, 208, 209] and CLPSO on the investigation of protein structure prediction. The HPSODE algorithm was found superior results over real protein sequences.

8.2 Future Research Directions

In this book different new methodologies have been advised on various types of protein sequences. However, still, there is a huge scope for future work for all of the leads presented in this book. The study may be extended in a number of ways, some of which are discussed below.

- A few number of landscape features have been investigated for protein instances on AB off-lattice model in this book. A preliminary study on the properties of local optima for protein landscape structure has been done in our work. The future work can be extended to evaluate the expected time to reach local optima, the distance between local optima and the expected probability of reaching the optima. Other important landscape features that have not been investigated include separability i.e. the level of variable interdependency in a problem affects fitness landscape as well as the performance of an algorithm. Finding ways of estimating separability in a function in a landscape structure is an area of future research.
- Fitness landscape analysis techniques are part of characterizing the features of a problem. To this end, automatic selection of algorithms for optimization is a potential task for future work for protein structure prediction. The possible framework for such system includes population-based algorithm portfolios where multiple algorithms are run in parallel with an interaction between them.
- Multi-objective optimization has become a subject of interest among the researchers who are working on optimization. The AB off-lattice model corresponds to an intra-molecular energy function which is the sum of bending potential energy totally based on angles and non-bonded interactions based on distance calculations. There is a scope of applying multi-objective optimization to understand the nature of the two energy components resulting minimization of the total energy function.
- In this book, we present an empirical study on metaheuristic based protein structure prediction and our investigation reveals better results on various protein instances. However, mathematical analysis of the dynamics of the proposed techniques serves as a promising future research.
- An interesting problem for future research is to associate the choice of potential cost or energy function which provides more molecular interactions among the amino acids. Thus, the CHARMM energy function, known as Chemistry at HARvard Macromolecular Mechanism can be considered as energy model for protein structure prediction. CHARMM uses potential functions that approximate the total potential as a sum of bond stretching, bond bending, bond twisting, planner bonds plus the non-bonded van der waals and electrostatic interactions.
- A more extensive real protein sequences with longer length (> 100) can be considered compared to artificial protein sequences for predicting the protein structure.

References

1. Honig, B.: Protein folding: from the levinthal paradox to structure prediction. J. Mol. Biol. **293**(2), 283–293 (1999)
2. Hendy, H., Khalifa, W., Roushdy, M., Salem, A.B.: A study of intelligent techniques for protein secondary structure prediction. Int. J. Inf. Mod. Anal. **4**(1), 3–12 (2015)
3. Freitas, A.A., Wieser, D.C., Apweiler, R.: On the importance of comprehensible classification models for protein function prediction. IEEE/ACM Trans. Comput. Biol. Bioinform. **7**(1), 172–182 (2010)
4. Stenson, P.D., Mort, M., Ball, E.V., Shaw, K., Phillips, A.D., Cooper, D.N.: The human gene mutation database: building a comprehensive mutation repository for clinical and molecular genetics, diagnostic testing and personalized genomic medicine. Hum. Genet. **133**(1), 1–9 (2014)
5. Soto, C.: Unfolding the role of protein misfolding in neurodegenerative diseases. Nat. Rev. Neurosci. **4**(1), 49–60 (2003)
6. Soto, C., Estrada, L.D.: Protein misfolding and neurodegeneration. Arch. Neurol. **65**(2), 184–189 (2008)
7. Uversky, V.N.: Intrinsic disorder in proteins associated with neurodegenerative diseases. In: Ovádi, J., Orosz, F. (eds.) Protein Folding and Misfolding: Neurodegenerative Diseases, pp. 21–75. Springer, Netherlands, Dordrecht (2009)
8. Neudecker, P., Robustelli, P., Cavalli, A., Walsh, P., Lundström, P., Zarrine-Afsar, A., Sharpe, S., Vendruscolo, M., Kay, L.E.: Structure of an intermediate state in protein folding and aggregation. Science **336**(6079), 362–366 (2012)
9. Comellas, G., Rienstra, C.: Protein structure determination by magic-angle spinning solid-state nmr, and insights into the formation, structure, and stability of amyloid fibrils. Annu. Rev. Biophys. **42**(1), 515–536 (2013)
10. Bagaria, A., Jaravine, V., Gntert, P.: Estimating structure quality trends in the protein data bank by equivalent resolution. Comput. Biol. Chem. **46**, 8–15 (2013)
11. Sousa, S.F., Fernandes, P.A., Ramos, M.J.: Protein-ligand docking: current status and future challenges. Proteins **65**(1), 15–26 (2006)
12. Venkatesan, A., Gopal, J., Gollapalli, S., Karthikeyan, K.: Computational approach for protein structure prediction. Healthc. Inform. Res. **19**(2), 137–147 (2013)
13. Protein sequence & structure information. http://www.rcsb.org/pdb/statistics/holdings.do. Accessed Jan 2017
14. Kim, S.Y., Lee, S.B., Lee, J.: Structure optimization by conformational spaceannealing in an off-lattice protein model. Phys. Rev. E **72**, 011916–6 (2005)

© Springer International Publishing AG 2018
N. D. Jana et al., *A Metaheuristic Approach to Protein Structure Prediction*, Emergence, Complexity and Computation 31,
https://doi.org/10.1007/978-3-319-74775-0

15. Floudas, C.A.: Computational methods in protein structure prediction. Biotechnol. Bioeng. **97**(2), 207–213 (2007)
16. Unger, R.: The genetic algorithm approach to protein structure prediction. In: Johnston, R.L. (ed.) Applications of Evolutionary Computation in Chemistry, pp. 153–175. Springer, Berlin (2004)
17. Bonneau, R., Baker, D.: Ab initio protein structure prediction: progress and prospects. Annu. Rev. Biophys. Biomol. Struct. **30**(1), 173–189 (2001). PMID: 11340057
18. Anfinsen, C.B.: Principles that govern the folding of protein chains. Science **181**, 223–230 (1973)
19. Bryant, S.H., Lawrence, C.E.: An empirical energy function for threading protein sequence through the folding motif. Proteins Struct. Funct. Bioinform. **16**(1), 92–112 (1993)
20. Dill, K.A.: Theory for the folding and stability of globular proteins. Biochemistry **24**(6), 1501–1509 (1985)
21. Stillinger, F.H., Head-Gordon, T., Hirshfel, C.L.: Toy model for protein folding. Phys. Rev. **48**, 1469–1477 (1993)
22. Dill, A.K., Bromberg, S., Yue, K., Fiebig, K.M., Yee, D.P.: Thomas, P.D., chan, H.S.: Principle of protein folding: a perspective from simple exact models. Protein Sci. **4**(4), 561–602 (1995)
23. Unger, R., Moult, J.: Finding the lowest free energy conformation of a protein is an np-hard problem: proof and implications. Bull. Math. Biol. **55**(6), 1183–1198 (1993)
24. Atkins, J., Hart, W.E.: On the intractability of protein folding with a finite alphabet of amino acids. Algorithmica **25**(2), 279–294 (1999)
25. Berger, B., Leighton, T.: Protein folding in the hydrophobic-hydrophilic (hp) is np-complete, In: Proceedings of the Second Annual International Conference on Computational Molecular Biology, RECOMB '98, New York, NY, USA, pp. 30–39. ACM (1998)
26. Crescenzi, P., Goldman, D., Papadimitriou, C., Piccolboni, A., Yannakakis, M.: On the complexity of protein folding. J. Comput. Biol. **5**(3), 423–465 (1998)
27. Leach, A.R.: Molecular Modeling Principles and Applications. Pearson, Harlow (2010)
28. Yan, L.: Conditional graphical models for protein structure prediction. Ph.D. thesis, Carnegie Mellon University (2006)
29. Li, Z., Scheraga, H.A.: Monte carlo-minimization approach to the multiple-minima problem in protein folding. Proc. Natl. Acad. Sci. **84**(19), 6611–6615 (1987)
30. Li, Z., Scheraga, H.A.: Structure and free energy of complex thermodynamic systems. J. Mol. Struct. (THEOCHEM) **179**(1), 333–352 (1988)
31. Hoque, M.T., Chetty, M., Dooley, L.S.: Significance of hybrid evolutionary computation for ab initio protein folding prediction. In: Abraham, A., Grosan, C., Ishibuchi, H. (eds.) Hybrid Evolutionary Algorithms, pp. 241–268. Springer, Berlin (2007)
32. Hodgman, C., French, A., Westhead, D.: Instant Notes in Bioinformatics, 2nd edn. Taylor & Francis Inc, Bristol (2009)
33. Corne, D.W., Fogel, G.B.: An introduction to bioinformatics for computer scientists. In: Fogel, G.B., Corne, D.W. (eds.) Evolutionary Computation in Bioinformatics. The Morgan Kaufmann Series in Artificial Intelligence, pp. 3–18. Morgan Kaufmann, San Francisco (2003)
34. Cohen, J.: Bioinformatics-an introduction for computer scientists. ACM Comput. Surv. **36**, 122–158 (2004)
35. Guex, N., Peitsch, M.C.: Protein structure: comparative protein modelling and visualisation. http://swissmodel.expasy.org/course/course-index.htm (2007)
36. Jones, D.T., Miller, R.T., Thornton, J.M.: Successful protein fold recognition by optimal sequence threading validated by rigorous blind testing. Proteins Struct. Func. Bioinform. **23**(3), 387–397 (1995)
37. Krogh, A., Brown, M., Mian, I., Sjlander, K., Haussler, D.: Hidden markov models in computational biology. J. Mol. Biol. **235**(5), 1501–1531 (1994)
38. Chou, P.Y., Fasman, G.D.: Prediction of protein conformation. Biochemistry **13**(2), 222–245 (1974). PMID: 4358940
39. Rahman, A., Zomaya, A.Y.: An overview of protein-folding techniques: issues and perspectives. Int. J. Bioinform. Res. Appl. **1**(1), 121–143 (2005). PMID: 18048125

40. Snchez, R., Sali, A.: Large-scale protein structure modeling of the saccharomyces cerevisiae genome. Proc. Natl. Acad. Sci. **95**(23), 13597–13602 (1998)
41. Jones, D.T.: Genthreader: an efficient and reliable protein fold recognition method for genomic sequences 1. J. Mol. Biol. **287**(4), 797–815 (1999)
42. Samudrala, R., Xia, Y., Levitt, M., Huang, E.: A combined approach for ab initio construction of low resolution protein tertiary structures from sequence. Pac. Symp. Biocomput. 505–516 (1999)
43. Dill, K.A.: Theory for the folding and stability of globular proteins. Biochemestry **24**(6), 1501–1509 (1985)
44. Li, H., Helling, R., Tang, C., Wingreen, N.: Emergence of preferred structures in a simple model of protein folding. Science **273**(5275), 666–669 (1996)
45. Irback, A., Peterson, C., Potthast, F., Sommelius, O.: Local interactions and protein folding: a 3d off-lattice approach. Chem. Phys. **107**(1), 273–282 (1997)
46. Bandaru, S., Deb, K.: Metaheuristic techniques. In: Sengupta, R., Gupta, A., Dutta, J. (eds.) Decision Sciences: Theory and Practice, pp. 693–750. Taylor & Francis Group (2016)
47. Storn, R., Price, K.: Differential evolution: a simple and efficient heuristic for global optimization over continuous spaces. J. Glob. Optim. **11**(4), 341359 (1997)
48. Das, S., Suganthan, P.N.: Differential evolution: A survey of the state-of-the-art. IEEE Trans. Evol. Comput. **15**(1), 4–31 (2011)
49. Das, S., Mullick, S.S., Suganthan, P.: Recent advances in differential evolution an updated survey. Swarm Evol. Comput. **27**, 1–30 (2016)
50. Price, K., Storn, R.M., Lampinen, J.A.: Differential Evolution: A Practical Approach to Global Optimization. Natural Computing Series. Springer-Verlag, New York Inc., Secaucus (2005)
51. Storn, R., Price, K.: Differential evolution - a simple and efficient heuristic for global optimization over continuous spaces. J. Glob. Optim. **11**, 341–359 (1997)
52. Simon, D.: Biogeography- based optimization. IEEE Trans. Evol. Comput. **12**(6), 702–713 (2008)
53. Beni, G., Wang, J.: Swarm intelligence in cellular robotic systems. In: Dario, P., Sandini, G., Aebischer, P. (eds.) Robots and Biological Systems: Towards a New Bionics?, pp. 703–712. Springer, Berlin (1993)
54. Kennedy, J., Eberhart, R.: Particle swarm optimization. In: IEEE International Conference on Neural Networks (ICNN95), pp. 1942–1948. Perth, IEEE Press, Australia (1995)
55. Eberhart, R., Kennedy, J.: A new optimizer using particle swarm theory. In: Proceedings of the Sixth International Symposium on Micro Machine and Human Science, MHS '95, October 1995, pp. 39–43 (1995)
56. Eberhart, R.C., Shi, Y.: Particle swarm optimization: developments, applications and resources. In: Proceedings of the 2001 Congress on Evolutionary Computation (IEEE Cat. No.01TH8546), vol. 1, pp. 81–86 (2001)
57. Shi, Y., Eberhart, R.: A modified particle swarm optimizer. In: 1998 IEEE International Conference on Evolutionary Computation Proceedings. IEEE World Congress on Computational Intelligence (Cat. No.98TH8360), May 1998, pp. 69–73 (1998)
58. Xie, X.-F., Zhang, W.-J., Yang, Z.-L.: Adaptive particle swarm optimization on individual level. In: 6th International Conference on Signal Processing, August 2002, vol. 2, pp. 1215–1218 (2002)
59. Pham, D., Castellani, M.: The bees algorithmmodelling foraging behaviour to solve continuous optimisation problems. J. Bio.-Inspired Comput. **223**(12), 2919–2938 (2009)
60. Geem, Z.W., Kim, J.H., Loganathan, G.V.: A new heuristic optimization algorithm: harmony search. Simulation **76**(2), 60–68 (2001)
61. Parpinelli, R.S., Lopes, H.S.: An ecology-based evolutionary algorithm applied to the 2d-ab off-lattice protein structure prediction problem. In: 2013 Brazilian Conference on Intelligent Systems, October 2013, pp. 64–69 (2013)
62. Olson, B., De Jong, K., Shehu, A.: Off-lattice protein structure prediction with homologous crossover. In: Proceedings of the 15th Annual Conference on Genetic and Evolutionary Computation, GECCO '13, New York, NY, USA, pp. 287–294. ACM, 2013

63. Bošković, B., Brest, J.: Differential evolution for protein folding optimization based on a three-dimensional ab off-lattice model. J. Mol. Model. **22**(10), 252 (2016)

64. Brest, J., Greiner, S., Boskovic, B., Mernik, M., Zumer, V.: Self-adapting control parameters in differential evolution: a comparative study on numerical benchmark problems. IEEE Trans. Evol. Comput. **10**, 646–657 (2006)

65. Kalegari, D.H., Lopes, H.S.: An improved parallel differential evolution approach for protein structure prediction using both 2d and 3d off-lattice models. In: IEEE Symposium on Differential Evolution (SDE), Singapore, pp. 143–150 (2013)

66. Kalegari, D.H., Lopes, H.S.: A differential evolution approach for protein structure optimization using a 2d off-lattice model. J. Bio.-Inspired Comput. **2**(3), 242–250 (2010)

67. Sar, E., Acharyya, S.: Genetic algorithm variants in predicting protein structure. In: 2014 International Conference on Communication and Signal Processing, April 2014, pp. 321–325 (2014)

68. Fan, J., Duan, H., Xie, G., Shi, H.: Improved biogeography-based optimization approach to secondary protein prediction. In: 2014 International Joint Conference on Neural Networks (IJCNN), July 2014, pp. 4223–4228 (2014)

69. Cheng-yuan, L., Yan-rui, D., Wen-bo, X.: Multiple-layer quantum-behaved particle swarm optimization and toy model for protein structure prediction. In: 2010 Ninth International Symposium on Distributed Computing and Applications to Business, Engineering and Science, Aug 2010, pp. 92–96 (2010)

70. Liu, J., Wang, L., He, L., Shi, F.: Analysis of toy model for protein folding based on particle swarm optimization algorithm. In: Wang, L., Chen, K., Ong, Y.S. (eds.) Advances in Natural Computation: First International Conference, ICNC 2005, Changsha, China, 27–29 Aug 2005, Proceedings, Part III, Berlin, Heidelberg, pp. 636–645. Springer, Berlin, Heidelberg (2005)

71. Zhu, H., Pu, C., Lin, X., Gu, J., Zhang, S., Su, M.: Protein structure prediction with epso in toy model. In: 2009 Second International Conference on Intelligent Networks and Intelligent Systems, November 2009, pp. 673–676 (2009)

72. Zhang, X., Li, T.: Improved particle swarm optimization algorithm for 2d protein folding prediction. In: 2007 1st International Conference on Bioinformatics and Biomedical Engineering, July 2007, pp. 53–56 (2007)

73. Chen, X., Lv, M.W., Zhao, L.H., Zhang, X.D.: An improved particle swarm optimization for protein folding prediction. Int. J. Inf. Eng. Electron. Bus. **3**(1), 1–8 (2011)

74. Li, B., Li, Y., Gong, L.: Protein secondary structure optimization using an improved artificial bee colony algorithm based on ab off-lattice model. Eng. Appl. Artif. Intell. **27**, 70–79 (2014)

75. Wang, Y., Guo, G.D., Chen, L.F.: Chaotic artificial bee colony algorithm: a new approach to the problem of minimization of energy of the 3d protein structure. J. Mol. Biol. **47**(6), 894–900 (2013)

76. Li, B., Chiong, R., Lin, M.: A balance-evolution artificial bee colony algorithm for protein structure optimization based on a three-dimensional ab off-lattice model. Comput. Biol. Chem. **54**, 1–12 (2015)

77. Li, B., Lin, M., Liu, Q., Li, Y., Zhou, C.: Protein folding optimization based on 3d off-lattice model via an improved artificial bee colony algorithm. J. Mol. Model. **21**(261), 1–15 (2015)

78. Parainelli, R.S., Benitez, C.M.V., Lopes, H.S.: Performance analysis of swarm intelligence algorithms for the 3d-ab off-lattice protein folding problem. J. Mult.-Valued Log. Soft Comput. **22**, 267–286 (2014)

79. Lin, X., Zhu, H.: Structure optimization by an improved tabu search in the ab off-lattice protein model. In: 2008 First International Conference on Intelligent Networks and Intelligent Systems, November 2008, pp. 123–126 (2008)

80. Lin, X., Zhang, X., Zhou, F.: Protein structure prediction with local adjust tabu search algorithm. BMC Bioinform. **15**(15), S1 (2014)

81. Liu, J., Sun, Y., Li, G., Song, B., Huang, W.: Heuristic-based tabu search algorithm for folding two dimensional ab off-lattice model proteins. Comput. Biol. Chem. **47**, 142–148 (2013)

82. Scalabrin, M.H., Parpinelli, R.S., Benítez, C.M.V., Lopes, H.S.: Population-based harmony search using gpu applied to protein structure prediction. Int. J. Comput. Sci. Eng. **9**, 106–118 (2014)

83. Lin, X., Zhang, X., Zhou, F.: Protein structure prediction with local adjust tabu search algorithm. BMC Bioinform. **15**, S1 (2014)

84. Mansour, R.F.: Applying an evolutionary algorithm for protein structure prediction. Am. J. Bioinform. Res. **1**(1), 18–23 (2011)

85. Zhang, X., Wang, T., Luo, H., Yang, J.Y., Deng, Y., Tang, J., Yang, M.Q.: 3d protein structure prediction with genetic tabu search algorithm. BMC Syst. Biol. **4**(1), S6 (2010)

86. Zhou, C., Hou, C., Wei, X., Zhang, Q.: Improved hybrid optimization algorithm for 3d protein structure prediction. J. Mol. Model. **20**(7), 2289 (2014)

87. Lin, X., Zhang, X.: Protein folding structure optimization based on gapso algorithm in the off-lattice model. In: 2014 IEEE International Conference on Bioinformatics and Biomedicine (BIBM), November 2014, pp. 43–49 (2014)

88. Malan, K.M., Engelbrecht, A.P.: A survey of techniques for characterising fitness landscapes and some possible ways forward. Inf. Sci. **241**, 148–163 (2013)

89. McClymont, K.: Recent advances in problem understanding: Changes in the landscape a year on. In: Proceedings of the 15th Annual Conference Companion on Genetic and Evolutionary Computation, GECCO '13 Companion, New York, NY, USA, pp. 1071–1078. ACM (2013)

90. Jana, N.D., Sil, J., Das, S.: Continuous fitness landscape analysis using a chaos-based random walk algorithm. Soft Comput. 1–28 (2016)

91. Jana, N.D., Sil, J., Das, S.: Selection of appropriate metaheuristic algorithms for protein structure prediction in AB off-lattice model: a perspective from fitness landscape analysis. Inf. Sci. **391392**, 28–64 (2017)

92. Jana, N.D., Sil, J.: Lévy distributed parameter control in differential evolution for numerical optimization. Nat. Comput. **15**(3), 371–384 (2016)

93. Jana, N.D., Sil, J.: Hybrid particle swarm optimization technique for protein structure prediction using 2d off-lattice model. In: Panigrahi, B.K., Suganthan, P.N., Das, S., Dash, S.S. (eds.) Swarm, Evolutionary, and Memetic Computing: 4th International Conference, SEMCCO 2013, Chennai, India, 19–21 December 2013, Proceedings, Part II, pp. 193–204. Springer International Publishing, Cham (2013)

94. Jana, N.D., Sil, J., Das, S.: Improved bees algorithm for protein structure prediction using ab off-lattice model. In: Matouek, R. (ed.) Mendel 2015. Advances in Intelligent Systems and Computing, pp. 39–52. Springer International Publishing, Berlin (2015)

95. Jana, N.D., Sil, J., Das, S.: Protein structure optimization in 3d ab off-lattice model using bbo with chaotic mutation. In: The 4th ACM IKDD Conferences on Data Sciences (CODS 2017). ACM (2017) (Accepted)

96. Jana, N.D., Sil, J., Das, S.: An improved harmony search algorithm for protein structure prediction using 3d off-lattice model. In: Del Ser, J. (ed.) Harmony Search Algorithm: Proceedings of the 3rd International Conference on Harmony Search Algorithm (ICHSA 2017), Singapore, pp. 304–314. Springer, Singapore (2017)

97. Jana, N.D., Sil, J.: Interleaving of particle swarm optimization and differential evolution algorithm for global optimization. Int. J. Comput. Appl. **38**(2–3), 116–133 (2016)

98. Blum, C., Li, X.: Swarm intelligence in optimization. In: Blum, C., Merkle, D. (eds.) Swarm Intelligence. Natural Computing Series, pp. 43–85. Springer, Berlin (2008)

99. Jong, K.D.: Parameter setting in eas: a 30 year perspective. In: Lobo, F.G., Lima, C.F., Michalewicz, Z. (eds.) Parameter Setting in Evolutionary Algorithms. Studies in Computational Intelligence, vol. 54, pp. 1–18. Springer, Berlin (2005)

100. Wolpert, D., Macready, W.: No free lunch theorems for optimization. IEEE Trans. Evol. Comput. **1**(1), 67–82 (1997)

101. Reeves, C.R., Rowe, J.E.: Genetic Algorithms - Principles and Perspectives: A Guide to GA Theory. Kluwer Academic Publishers, Norwell (2002)

102. Merz, P., Freisleben, B.: Fitness landscape analysis and memetic algorithms for the quadratic assignment problem. IEEE Trans. Evol. Comput. **4**(4), 337–352 (2000)

103. Naudts, B., Kallel, L.: A comparison of predictive measures of problem difficulty in evolutionary algorithms. IEEE Trans. Evol. Comput. **4**(1), 1–15 (2000)

104. Jones, T., Forrest, S.: Fitness distance correlation as a measure of problem difficulty for genetic algorithms. In: Sixth International Conference on Genetic Algorithms, pp. 184–192 (1995)
105. Rose, H., Ebeling, W., Asselmeyer, T.: The density of states - a measure of the difficulty of optimisation problems. In: 4th International Conference on Parallel Problem Solving from Nature, pp. 208–217 (1996)
106. Lunacek, M., Whitley, D.: The dispersion metric and the cma evolution strategy. In: 8th Annual Conference on Genetic and Evolutionary Computation, pp. 477–484 (2006)
107. Mersmann, O., Bischl, B., Trautmann, H., Preuss, M., Weihs, C., Rudolph, G.: Exploratory landscape analysis. In: 13th Annual Conference on Genetic and Evolutionary Computation (GECCO'11), pp. 829–836 (2011)
108. Tavares, J., Pereira, F.B., Costa, E.: Multidimensional knapsack problem: a fitness landscape analysis. IEEE Trans. Syst. Man Cybern. Part B Cybern. **38**(3), 604–616 (2008)
109. Vassilev, V.K., Fogarty, T.C., Miller, J.F.: Information characteristics and the structure of landscapes. Evol. Comput. **8**(1), 31–60 (2000)
110. Vassilev, V.K., Fogarty, T.C., Miller, J.E.: Smoothness, ruggedness and neutrality of fitness landscapes: from theory to application. Advances in Evolutionary Computing. Natural Computing Series, pp. 3–44. Springer, Berlin (2003)
111. Munoz, M.A., Kirley, M., Halgamuge, S.: Landscape characterization of numerical optimization problems using biased scattered data. In: IEEE Congress on Evolutionary Computation (CEC'12), pp. 1–8 (2012)
112. Vanneschi, L., Clergue, M., Collard, P., Tomassini, M., Vérel, S.: Fitness clouds and problem hardness in genetic programming. In: Deb, K., Poli, R., Banzhaf, W., Beyer, H.-G., Burke, E., Darwen, P., Dasgupta, D., Floreano, D., Foster, J., Harman, M., Holland, O., Lanzi, P.L., Spector, L., Tettamanzi, A., Thierens, D., Tyrrell, A. (eds.) Genetic and Evolutionary Computation – GECCO-2004, Part II, Seattle, WA, USA. Lecture Notes in Computer Science, vol. 3103, pp. 690–701. Springer (2004)
113. Lu, G., Li, J., Yao, X.: Fitness-probability cloud and a measure of problem hardness for evolutionary algorithms. In: Merz, P., Hao, J.-K. (eds.) Proceedings of Evolutionary Computation in Combinatorial Optimization: 11th European Conference, EvoCOP 2011, Torino, Italy, 27–29 April, 2011, (Berlin, Heidelberg), pp. 108–117. Springer, Berlin, Heidelberg (2011)
114. Malan, K.M., Engelbrecht, A.P.: Fitness landscape analysis for metaheuristic performance prediction. In: Richter, H., Engelbrecht, A.P. (eds.) Recent Advances in the Theory and Application of Fitness Landscapes. Emergence, Complexity and Computation, vol. 6, pp. 103–132. Springer, Berlin (2014)
115. Malan, K.M., Engelbrecht, A.P.: A progressive random walk algorithm for sampling continuous fitness landscapes. In: IEEE Congress on Evolutionary Computation (CEC'14), pp. 2507–2514 (2014)
116. Pearson, K.: The problem of the random walk. Nature **72**(1), 294, 318, 342 (1905)
117. Huang, S.Y., Zou, X.W., Jin, Z.Z.: Directed random walks in continuous space. Phys. Rev. E **65**(5), 052105 (2002)
118. Morgan, R., Gallagher, M.: Sampling techniques and distance metrics in high dimensional continuous landscape analysis: limitations and improvements. IEEE Trans. Evol. Comput. **18**(3), 456–461 (2014)
119. Malan, K.M., Engelbrecht, A.P.: Quantifying ruggedness of continuous landscapes using entropy. In: IEEE Congress on Evolutionary Computation (CEC'09), pp. 1440–1447 (2009)
120. Reidys, C.M., Stadler, P.F.: Neutrality in fitness landscapes. Appl. Math. Comput. **117**(2–3), 321–350 (2001)
121. Mohammad, H.T.-N., Bennett, A.P.: On the landscape of combinatorial optimization problems. IEEE Trans. Evol. Comput. **18**(3), 420–434 (2014)
122. Morgan, R., Gallagher, M.: Length scale for characterising continuous optimization problems. In: 12th International Conference on Parallel Problem Solving from Nature - Part I, pp. 407–416 (2012)
123. Munoz, M., Kirley, M., Halgamuge, S.: Exploratory landscape analysis of continuous space optimization problems using information content. IEEE Trans. Evol. Comput. **19**(1), 74–87 (2014)

124. Iba, T., Shimonishi, K.: The origin of diversity: thinking with chaotic walk. In: Proceedings of the Eighth International Conference on Complex Systems, pp. 447–461 (2011)

125. Tavazoei, M.S., Haeri, M.: Comparison of different one-dimensional maps as chaotic search pattern in chaos optimization algorithms. Appl. Math. Comput. **187**, 1076–1085 (2007)

126. Yuan, X., Dai, X., Wu, L.: A mutative-scale pseudo-parallel chaos optimization algorithm. Soft Comput. **19**, 1215–1227 (2015)

127. May, R.M.: Simple mathematical models with very complicated dynamics. Nature **261**, 459–467 (1976)

128. Peitgen, H., Jurgens, H., Saupe, D.: Chaos and Fractals. Springer, Berlin (1992)

129. Steer, K.C.B., Wirth, A., Halgamuge, S.K.: Information theoretic classification of problems for metaheuristics. In: Li, X., Kirley, M., Zhang, M., Green, D., Ciesielski, V., Abbass, H., Michalewicz, Z., Hendtlass, T., Deb, K., Tan, K.C., Branke, J., Shi, Y. (eds.) Proceedings of Simulated Evolution and Learning: 7th International Conference, SEAL 2008, Melbourne, Australia, 7–10 December 2008, Berlin, Heidelberg, pp. 319–328. Springer, Berlin, Heidelberg (2008)

130. Liang, J.J., Qu, B.Y., Suganthan, P.N., Hernandez-Diaz, A.G.: Problem definitions and evaluation criteria for the cec 2013 special session on real-parameter optimization. Technical Report, DAMTP 2000/NA10, Nanyang Technological University, Singapore (2013)

131. Caraffini, F., Neri, F., Picinali, L.: An analysis on separability for memetic computing automatic design. Inf. Sci. **265**, 1–22 (2014)

132. Venkatesan, A., et al.: Computational approach for protein structure prediction. Healthc. Inform. Res. **19**(2), 137–147 (2013)

133. Lozano, M., Molina, D., Herrera, F.: Editorial scalability of evolutionary algorithms and other metaheuristics for large-scale continuous optimization problems. Soft Comput. **15**(11), 2085–2087 (2011)

134. Dorn, M., Silva, M.B., Buriol, L.S., Lamb, L.C.: Three-dimensional protein structure prediction: methods and computational strategies. Comput. Biol. Chem. **53**, 251–276 (2014)

135. Parejo, J.A., Ruiz-Cortés, A., Lozano, S., Fernandez, P.: Metaheuristic optimization frameworks: a survey and benchmarking. Soft Comput. **16**(3), 527–561 (2012)

136. Boussad, I., Lepagnot, J., Siarry, P.: A survey on optimization metaheuristics. Inf. Sci. **237**, 82–117 (2013)

137. Sörensen, K.: Metaheuristicsthe metaphor exposed. Int. Trans. Oper. Res. **22**(1), 3–18 (2015)

138. Hutter, F., Hoos, H.H., Leyton-Brown, K., Stützle, T.: Paramils: an automatic algorithm configuration framework. J. Artif. Int. Res. **36**, 267–306 (2009)

139. Caamao, P., Bellas, F., Becerra, J.A., Duro, R.J.: Evolutionary algorithm characterization in real parameter optimization problems. Appl. Soft Comput. **13**(4), 1902–1921 (2013)

140. Blum, C., Puchinger, J., Raidl, G.R., Roli, A.: Hybrid metaheuristics in combinatorial optimization: a survey. Appl. Soft Comput. **11**(6), 4135–4151 (2011)

141. Weise, T., Zapf, M., Chiong, R., Nebro, A.J.: Why is optimization difficult? pp. 1–50. Springer, Berlin (2009)

142. Muñoz, M.A., Kirley, M., Halgamuge, S.K.: The algorithm selection problem on the continuous optimization domain, pp. 75–89. Springer, Berlin (2013)

143. Kotthoff, L.: Algorithm selection for combinatorial search problems: a survey. AI Mag. **35**(3), 48–60 (2014)

144. Kanda, J., Carvalho, A., Hruschka, E., Soares, C.: Selection of algorithms to solve traveling salesman problems using meta-learning. Int. J. Hybrid Intell. Syst. **8**(3), 117–128 (2011)

145. Muoz, M.A., Sun, Y., Kirley, M., Halgamuge, S.K.: Algorithm selection for black-box continuous optimization problems: a survey on methods and challenges. Inf. Sci. **317**, 224–245 (2015)

146. Eiben, A., Hinterding, R., Michalewicz, Z.: Parameter control in evolutionary algorithms. IEEE Trans. Evol. Comput. **3**(2), 124–141 (1999)

147. Eiben, A., Smit, S.: Parameter tuning for configuring and analyzing evolutionary algorithms. Swarm Evol. Comput. **1**(1), 19–31 (2011)

148. Pedersen, M.: Tuning & simplifying heuristical optimization. Ph.D. Dissertation, University of Southampton (2009)
149. Smith, J.E., Fogarty, T.C.: Operator and parameter adaptation in genetic algorithms. Soft Comput. **1**(2), 81–87 (1997)
150. Wolpert, D.H., Macready, W.: No free lunch theorems for search. Technical Report SFI-TR-95-02-010, Nanyang Technological University, February 1995
151. Wolpert, D.H., Macready, W.G.: No free lunch theorems for optimization. IEEE Trans. Evol. Comput. **1**(1), 67–82 (1997)
152. Muñoz, M.A., Kirley, M., Halgamuge, S.K.: A meta-learning prediction model of algorithm performance for continuous optimization problems, pp. 226–235. Springer, Berlin (2012)
153. Marín, J.: How landscape ruggedness influences the performance of real-coded algorithms: a comparative study. Soft Comput. **16**(4), 683–698 (2012)
154. Merz, P.: Advanced fitness landscape analysis and the performance of memetic algorithms. Evol. Comput. **12**(3), 303–325 (2004)
155. He, J., Reeves, C., Witt, C., Yao, X.: A note on problem difficulty measures in black-box optimization: classification, realizations and predictability. Evol. Comput. **15**(4), 435–443 (2007)
156. Pitzer, E., Affenzeller, M.: A comprehensive survey on fitness landscape analysis. In: Fodor, J., Klempous, R., Suárez Araujo, C.P. (eds.) Recent Advances in Intelligent Engineering Systems, pp. 161–191. Springer, Berlin (2012)
157. Sun, Y., Halgamuge, S.K., Kirley, M., Munoz, M.A.: On the selection of fitness landscape analysis metrics for continuous optimization problems. In: 7th International Conference on Information and Automation for Sustainability, pp. 1–6 (2014)
158. Davidor, Y.: Epistasis variance: suitability of a representation to genetic algorithms. Complex Syst. **4**, 369–383 (1990)
159. Tomassini, M., Vanneschi, L., Collard, P., Clergue, M.: A study of fitness distance correlation as a difficulty measure in genetic programming. Evol. Comput. **13**(2), 213–239 (2005)
160. Weinberger, E.: Correlated and uncorrelated fitness landscapes and how to tell the difference. Biol. Cybern. **63**(5), 325–336 (1990)
161. Sutton, A.M., Whitley, D., Lunacek, M., Howe, A.: Pso and multi-funnel landscapes: how cooperation might limit exploration. In: 8th Annual Conference on Genetic and Evolutionary Computation, pp. 75–82. ACM, New York (2006)
162. Bratley, P., Fox, B.L.: Implementing sobols quasirandom sequence generator. Assoc. Comput. Mach. **14**, 88–100 (1988)
163. Santana, R., Larraaga, P., Lozano, J.A.: Protein folding in simplified models with estimation of distribution algorithms. IEEE Trans. Evol. Comput. **12**(4), 418–438 (2008)
164. Islam, M., Chetty, M.: Clustered memetic algorithm with local heuristics for ab initio protein structure prediction. IEEE Trans. Evol. Comput. **17**(4), 558–576 (2013)
165. Boskovic, B., Brest, J.: Genetic algorithm with advanced mechanisma applied to the protein structure prediction in a hydrophobic-polar and cubic lattice. Appl. Soft Comput. **45**, 61–70 (2016)
166. Hansen, N., Auger, A., Finck, S., Ros, R.: Real-parameter black-box optimization benchmarking bbob-2010:experimental setup. Technical Report RR-7215, INRIA, France, September 2010
167. Goldberg, D.: Genetic Algorithms in Search. Optimization and Machine Learning. Longman Publishing Co., Boston, Addison-Wesley Professional (1989)
168. Karaboga, D., Basturk, B.: A powerful and efficient algorithm for numerical function optimization: artificial bee colony (abc) algorithm. J. Glob. Optim. **39**(3), 459–471 (2007)
169. Hansen, N., Ostermeier, A.: Completely derandomized self-adaptation in evolution strategies. Evol. Comput. **9**(2), 159–195 (2001)
170. Das, S., Konar, A., Chakraborty, U.K.: Two improved differential evolution schemes for faster global search. In: Proceedings of the 7th Annual Conference on Genetic and Evolutionary Computation, GECCO '05, New York, NY, USA, pp. 991–998. ACM (2005)

171. Pham, D.T., Castellani, M.: Benchmarking and comparison of nature-inspired population-based continuous optimisation algorithms. Soft Comput. **18**, 871–903 (2014)

172. Derrac, J., Garcia, S., Molina, D., Herrera, F.: A practical tutorial on the use of nonparametric statistical tests as a methodology for comparing evolutionary and swarm intelligence algorithms. Swarm Evol. Comput. **1**(1), 3–18 (2011)

173. Storn, R., Price, K.: Differential evolution: a simple and efficient adaptive scheme for global optimization over continuous spaces. Technical Report TR-95-012, ICSI, USA, March 1995

174. Črepinšek, M., Liu, S.-H., Mernik, M.: Exploration and exploitation in evolutionary algorithms: a survey. ACM Comput. Surv. **45**, 35:1–35:33 (2013)

175. Das, S., Konar, A., Chakraborty, U.K.: Improved differential evolution algorithms for handling noisy optimization problems. In: 2005 IEEE Congress on Evolutionary Computation, September 2005, vol. 2, pp. 1691–1698 (2005)

176. Chen, C.H., Lin, C.J., Lin, C.T.: Nonlinear system control using adaptive neural fuzzy networks based on a modified differential evolution. IEEE Trans. Syst. Man Cybern. Part C (Applications and Reviews) **39**, 459–473 (2009)

177. Abbass, H.A.: The self-adaptive pareto differential evolution algorithm. In: Proceedings of the 2002 Congress on Evolutionary Computation, 2002. CEC '02, May 2002, vol. 1, pp. 831–836 (2002)

178. Liu, J., Lampinen, J.: A fuzzy adaptive differential evolution algorithm. Soft Comput. **9**(6), 448–462 (2005)

179. Qin, A.K., Huang, V.L., Suganthan, P.N.: Differential evolution algorithm with strategy adaptation for global numerical optimization. IEEE Trans. Evol. Comput. **13**, 398–417 (2009)

180. Zaharie, D., Petcu, D.: Adaptive pareto differential evolution and its parallelization. In: Wyrzykowski, R., Dongarra, J., Paprzycki, M., Waśniewski, J. (eds.) Parallel Processing and Applied Mathematics: 5th International Conference, PPAM 2003, Czestochowa, Poland, September 7-10, 2003. Revised Papers, Berlin, Heidelberg, pp. 261–268. Springer, Berlin, Heidelberg (2004)

181. Mallipeddi, R., Suganthan, P., Pan, Q., Tasgetiren, M.: Differential evolution algorithm with ensemble of parameters and mutation strategies. Appl. Soft Comput. **11**(2), 1679–1696 (2011). The Impact of Soft Computing for the Progress of Artificial Intelligence

182. Thangaraj, R., Pant, M., Abraham, A.: A simple adaptive differential evolution algorithm. In: 2009 World Congress on Nature Biologically Inspired Computing (NaBIC), December 2009, pp. 457–462 (2009)

183. Levy, P.F.: The nut island effect. when good teams go wrong. Harv. Bus. Rev. **79**(3), 51–59 (2001)

184. Zolotarev, V.M.: One-Dimensional Stable Distributions. American Mathematical Society (1992)

185. Samoradnitsky, G., Taqqu, M.: Stable Non-Gaussian Random Processes: Stochastic Models with Infinite Variance. Chapman and Hall/CRC, Boca Raton (1994)

186. Johnson, N.L., Kotz, S., Balakrishnan, N.: Continuous Univariate Distributions. Wiley Series in Probability and Statistics. Wiley, New York (1994)

187. Suganthan, P.N., Hansen, N., Liang, J.J., Deb, K., Chen, Y.P., Auger, A., Tiwari, S.: Problem definitions and evaluation criteria for the cec 2005 special session on real-parameter optimization. Technical Repprt KanGal 2005005, Nanyang Technological University, Singapore & Kanpur Genrtic Algorithms Laboratory, IIT Kanpur, May 2005

188. Vesterstrom, J., Thomsen, R.: A comparative study of differential evolution, particle swarm optimization, and evolutionary algorithms on numerical benchmark problems. In: Proceedings of the 2004 Congress on Evolutionary Computation (IEEE Cat. No.04TH8753), June 2004, vol. 2, pp. 1980–1987 (2004)

189. Mount, D.W.: Bioinformatics: Sequence and Genome Analysis. Cold Spring Harbor Laboratory Press, Cold Spring Harbor (2001)

190. Kundu, R., Das, S., Mukherjee, R., Debchoudhury, S.: An improved particle swarm optimization with difference mean based perturbation. Neurocomputing **129**, 315–333 (2014)

191. Xiong, G., Shi, D., Duan, X.: Enhancing the performance of biogeography-based optimization using polyphyletic migration operator and orthogonal learning. Comput. Oper. Res. **41**, 125–139 (2014)
192. Ashrafi, S.M., Dariane, A.B.: Performance evaluation of an improved harmony search algorithm for numerical optimization: melody search (ms). Eng. Appl. Artif. Intell. **26**(4), 1301–1321 (2013)
193. Stillinger, F.H.: Collective aspects of protein folding illustrated by a toy model. Phys. Rev. E **52**(3), 2872–2877 (1995)
194. Saha, A., Datta, R., Deb, K.: Hybrid gradient projection based genetic algorithms for constrained optimization. In: 2010 IEEE Congress on Evolutionary Computation (CEC), July 2010, pp. 1–8 (2010)
195. El-Abd, M., Hassan, H., Anisa, M., Kamela, M.S., Elmasry, M.: Discrete cooperative particle swarm optimization for fpga placement. Appl. Soft Comput. **10**(4), 284–295 (2010)
196. Tsang, P.W.M., Yuena, T.Y.F., Situ, W.C.: Enhanced invariant matching of broken boundaries based on particle swarm optimization and the dynamic migrant principle. Appl. Soft Comput. **8**(5), 432–438 (2010)
197. Deb, K.: Optimization for Engineering Design: Algorithms and Examples. PHI Learning Pvt. Ltd., Delhi (2012)
198. Talbi, E.: A taxonomy of hybrid metaheuristics. J. Heuristic **8**(5), 541–564 (2002)
199. Črepinšek, M., Liu, S., Mernik, M.: Exploration and exploitation in evolutionary algorithms: a survey. ACM Comput. Surv. **45**(3), 1–33 (2013). Article 35
200. Clerc, M., Kennedy, J.: The particle swarm: explosion, stability and convergence in a multi-dimensional complex space. IEEE Trans. Evol. Comput. **6**(1), 58–73 (2002)
201. Noman, N., Iba, H.: Accelerating differential evolution using an adaptive local search. IEEE Trans. Evol. Comput. **12**(1), 107–125 (2008)
202. Pant, M., Thangaraj, R., Abraham, A.: Bde-pso: a new hybrid meta-heuristic for solving global optimization problems. New Math. Nat. Comput. **7**(3), 363–381 (2009)
203. Kim, P., Lee, J.: An integrated method of particle swarm optimization and differential evolution. J. Mech. Sci. Technol. **32**(23), 426–434 (2009)
204. Zheng, Q., Simon, D., Richter, H., Gao, Z.: Differential particle swarm evolution for robot control tuning. In: American Control Conference (ACC), pp. 5276–5281 (2014)
205. Back, T., Fogel, D., Michalewicz, Z.: Handbook of Evolutionary Computation. Oxford University Press, Oxford (1997)
206. Saha, A., Datta, R., Deb, K.: Hybrid gradient projection based genetic algorithms for constrained optimization. In: 2010 IEEE Congress on Evolutionary Computation (CEC), July 2010, pp. 1–8 (2010)
207. Fan, H., Lampinen, J.: A trigonometric mutation operation to differential evolution. J. Glob. Optim. **27**(1), 105–129 (2003)
208. Zhang, W.-J., Xie, X.-F.: Depso: hybrid particle swarm with differential evolution operator. IEEE International Conference on Systems Man and Cybernetics **4**, 3816–3821 (2003)
209. Chen, J., Xin, B., Peng, Z., Pan, F.: Statistical learning makes the hybridization of particle swarm and differential evolution more efficient- a novel hybrid optimizer. Sci. China Ser. Found. Inf. Sci. **52**(7), 1278–1282 (2009)
210. Liang, J.J., Qin, A.K., Suganthan, P.N., Baskar, S.: Comprehensive learning particle swarm optimizer for global optimization of multimodal functions. IEEE Trans. Evol. Comput. **10**, 281–295 (2006)

Printed in the United States
By Bookmasters